Telecommunication 4.0

Zhengmao Li

Telecommunication 4.0

Reinvention of the Communication Network

 Springer

Zhengmao Li
China Mobile
Beijing
China

ISBN 978-981-10-6300-8   ISBN 978-981-10-6301-5   (eBook)
DOI 10.1007/978-981-10-6301-5

Library of Congress Control Number: 2017949158

Original Chinese edition published by CITIC Press, China, 2016

Translation from the Chinese language edition: TONGXIN4.0: CHONGXIN FAMING TONGXIN-
WANG by Xiaodong Duan, Yachen Wang, Hao Zhang, Liang Geng, Yige Zhang, Qiao Fu, Jiayuan
Chen, Yuan Liu and Yuchen Wang, © CITIC Press, China 2016. All Rights Reserved.

Printed on acid-free paper

This Springer imprint is published by Springer Nature
The registered company is Springer Nature Singapore Pte Ltd.
The registered company address is: 152 Beach Road, #21-01/04 Gateway East, Singapore 189721, Singapore

Foreword I

The Chemistry between IT and CT

The development of communication networks is a mixed process of constant innovation, self-revolution and reinvention. After the evolution from analog to digital, and further to IP-based packet technology, communication network is now stepping into the era of telecommunication 4.0 supported by common Information Technology.

For more than a hundred years, the role of communication networks has shifted from a simple medium that solely connects people by delivering messages to a giant and sophisticated system that connects not only human beings but also "Things" across real and virtual spaces, significantly changing how people work, live and how the society runs in many aspects. These changes are non-stopping and communication technologies have been continuously advanced during the process of meeting and creating new demands. Looking at the history of communication technology development, the evolution from 1.0 to 3.0 era demonstrates a process of continuous accumulation and constant inspiration. In this development process, many significant breakthroughs have been made in key technologies of critical elements in the network, which also energizes the development of the human society. The emergence of the 4.0 era is expected as another significant milestone of the continuously-advancing communication technologies.

Information Technology (IT) and Communication Technology (CT) are regarded as two different technology areas before the rise of telecommunication 4.0 era. They essentially evolve independently in different communities. IT generally refers the wide range of technologies that derive from computer science. Essentially, CT shares the fundamental technology basis as in IT. However, CT, which is referred mostly by operators, requires high reliability, large scale and strong consistency. This results in a relatively closed industry with dedicated ecological system and product lines, which suffers from low flexibility, bad efficiency and high costs. In contrast, IT, which is referred mostly by Internet companies, is known for providing "best effort" services. They make use of the advanced computing technologies and common infrastructures, which has greatly improved the flexibility of the network and reduced the overall costs.

The explosive growth of data traffic and number of Internet applications has brought significant impact to traditional CT life cycle. Costs, efficiency and flexibility appear to become obvious bottlenecks. There is strong requirement for new technologies which could possibly break those constraints. IT, which advances all the way in line with Moore's Law, naturally becomes potential extra-power injected to CT thanks to its advanced technologies and mature communities. Introducing IT into CT can make up the weakness of communication industry whilst leverage the benefits of common hardware, low costs and high flexibility.

More importantly, as the cloud computing and big data technologies emerge with improved maturity and network bandwidth rapidly grows, it becomes feasible to introduce the latest IT in the communication field. The core of communication 4.0 is the deployment of universal computing, storage and network elements, software-defined communication functionalities, automatic OAM and centralized optimization of the network. In other words, the re-architected communication network is formed by software-defined elements whilst maintain the quality of CT. The comprehensive combination of CT and IT enables the overall coordination thorough capability exposure, elastic capacity and flexible architectural adjustment of the network. These capabilities provided by the new communication network make it possible for operators to define and develop their networks according to specific requirements, offering customized services to clients.

Nowadays communication network has already become the foundation of concept including Internet+, Internet of Things, Smart City and Intelligent life, etc. The emerging of communication 4.0 will have crucial effects upon Internet+, Industry 4.0 strategy and significantly influence the industrial chain.

First, let's talk about how it affects Internet+ strategy. This new type of industry, "Internet plus", with infinite possibilities, widely diversified fields and non-stop innovation, set new requirements for Internet infrastructure and corresponding operation. Traditional telecommunication network is generally considered as a walled garden. Facing the requirement of a more opened network, it is common for telecommunication network to introduce a layer of "open platform" to expose the network capabilities. This inevitably reduces the effectiveness and efficiency of the functionalities. Telecommunication 4.0 provides the openness at the level of the fundamental network elements. This extensive exposures greatly support and match up with "Internet Plus" industry requirements.

Second, telecommunication 4.0 has significant effect upon Industry 4.0. The Industry 4.0 strategy initiated by Germany triggered a booming industrial development in modern times. It is not mistaken to regard industry 4.0 as the fourth industrial revolution in human history, which characterizes the nature of intelligent manufacturing. It integrates information and communication technology with Cyber-Physical Systems (CPS), making the modern manufacturing industry more intelligent. The traditional telecommunication network is not designed for vertical applications and has to establish dedicated networks to meet customized requirements. Essentially, telecommunication 4.0 will make fundamental change to the industry. This is achieved by key enablers such as network slice technology, which improves the network flexibility and meets variety of requirements raised by vertical industries.

Finally, the chemistry of IT and CT will potentially reconstruct the industrial chain. Previously, the huge gap between IT and CT has set a barrier in between and it was extremely hard for players in IT industrial chain to step into the CT industry. Thus, IT was rarely introduced. However, since the infrastructure of CT network consists of universal IT hardware, new opportunities are created for IT industry. In turn, CT industry will also be energized by IT at the same time. Wide application of open-source software forces operators to transform from the old lifecycle of Purchasing, Construction, Maintenance and Selling to a "DevOps". This transformation will eventually cause another reconstruction of the industrial chain.

The chemistry brought by the alliance of IT industry and CT industry will be long lasting. This book looks back the way telecommunication network has developed alone with, explains the inevitability of telecommunication 4.0 era, illustrates the key technologies in telecommunication 4.0 era and discusses how new models, strategies and thoughts are developed under this new generation of network. In the meantime, this book makes some practical proposals for the participants in the telecommunication industry to embrace the telecommunication 4.0 era. At last, several reports from the leading organizations in the industry also discuss their views of telecommunication 4.0.

Beijing   Zhengmao Li

Foreword II

Telecommunication versus "Lego"

What does the network look like in the era of telecommunication 4.0? Let's start with something similar—Lego. Lego is so popular among both children and adults. But what makes Lego so special and attractive? Lego is a bunch of building blocks of different shapes. Let's call them the basic modules. These pieces can be assembled to build various models as you imagine. It might be a kind of animal, a piece of architecture, or even in more complex form—streets and cities. Since it's invention, Lego also upgrades and evolves. To date, many parents have already become fans of Lego since they bought the first set of Lego for their child. You will find nowadays that basic modules of Lego have more and more shapes and functionalities. But at the same time, they maintain high universality. Basic modules are always compatible with each other. This gives a massive space for creativity. The compatibility of basic modules never changes though Lego keep releasing new model in the market. To summarize, Lego is a good example of the creation of infinite number of models using universal building blocks. And it is very simple as follow,

1. Buy a bunch of basic modules
2. Gather all the basic modules you need for your own creation or according to the manual.
3. Make the most of your imagination and start assembling!

Fig. 1 The sketch of lego

If we compare the telecommunication network with Lego, the traditional telecommunication network is similar to a traditional toy of which the parts are fixed. Each elements of the telecommunication network is invariable as the parts of the toy. If the network functions need upgrades, new purchase or substitutions need to be carried out depending on the requirements. In telecommunication 4.0 era, the network is more like Lego and the construction of the network is similar to the process of assembling Lego as shown below,

1. Deploy basic general-purpose hardware resources
2. Coordinate the virtualized resources through the middle layer
3. Realize network functions using software

The basic general-purpose hardware resources act like the basic modules of Lego with strong diversity, versatility and availability. The intelligent middle layer, like the brain of a person, is able to coordinate necessary basic resources according to specific requirement, which also support the software-based upper-layer network functions. The network constructed in this way has the advantages of fast TTM, self-organization, self-restoration and elastic capacity. Hence, it is interesting to understand how telecommunication 4.0 networks could realize network functions with high flexibility and efficiency.

Beijing   Zhengmao Li

Preface I

Nokia is proud of its strong relationship with China Mobile. The author, Mr. Li Zhengmao, was inspired to think through some of the key issues that our industry faces during a visit to Finland in 2015. The results of Mr. Li's endeavors are outlined in this perceptive and insightful book.

Mr. Li summarizes the journey that the telecommunications industry has travelled, from analog, to digital, to IP, and predicts that future communication networks will combine both IT and telecommunications to become "telecommunication 4.0". The author elaborates on the key technologies that "telecommunication 4.0" will require, and reflects the demands and trends of future communications networks.

At Nokia we have risen to the challenges that "telecommunications 4.0" will bring. We also see a world where the network is growing in importance as data traffic increases, and the internet of things and the promise of 5G create new network requirements, new use cases, and new business models. The ten year work program that Bell Labs has undertaken shows what will be required, and we have worked on the technologies needed for this new world. To build on this momentum requires a scalable, responsive, programmable network with lower latency and faster speeds.

The integration of the IT world and the telcom world, as the author describes, will allow these requirements to be met. By combining new approaches and innovative technologies like Network Function Virtualization, Software-Defined Networking, and edge cloud interfaces, mobile operators will be able to re-imagine how connectivity can be used. At Nokia, we have not stopped there, we are looking further at new spectrum, and new business models, in order to ensure that the network is not merely an add-on, but an enabler—allowing operators, individuals and enterprises to make the most of the opportunities ahead of them.

At Nokia, we are inspired by the potential of these opportunities. Our vision is to expand the human possibilities of the connected world. We believe that this programmable world will have revolutionary consequences for us all, and that the network will be nothing less than the fabric of our connected lives. Nokia will continue to collaborate with China Mobile as we look to make this vision a reality. I urge you all to read Mr. Li's book, and think deeply about how you too can contribute to this important work.

Helsinki, Finland   Rajeev Suri
                                                          President and CEO, Nokia

Preface II

There are very few people who know the China's telecommunications industry better than Mr. Li Zhengmao. Mr. Li is the rare expert, who has both an academic background as well as having held a variety of senior positions in the largest telecommunications companies in China.

His engineering background and business acumen exceeds most in this highly competitive industry and he has seen tremendous change from a theoretical as well as an practical view.

The new perspectives used by the Mr. Li show his immense knowledge of the topics and sheds new light on the future of China's Telecommunications industry.

Mr. Li's description of the technology evolution and the Telecommunication 4.0 era is an early view of the industry will develop. By looking back at the history of the telecommunication industry, Mr. Li predicts future developments for Telecom Operators specifically and generally the Telecom Industry. Focusing on the future of the Telecommunication industry, the author outlines a blueprint, which will provide, in his opinion, a more open, converged and virtual path for the future.

I think that the era of Telecommunication 4.0 will impact the whole industry profoundly. It will bring more opportunities to companies in the CT/IT industry chain and trigger the restructuring of it. Internet companies and all telecom companies will continue to be interdependent on each other in order to compete in the market place. And China will accelerate the transformation of the industrial structure that drives one of the globe's greatest industries.

Mr. Li has made a great contribution to the telecommunications industry by writing this book. His thoughts and suggestions on a way forward will be studied and examined by academics and professionals alike.

USA

Craig Ehrlich
Chairman of GTI

Preface III

The emergence and development of a technology era do not merely rely on technical progress. The commercial environment and social civilization during the same period both affect technology. Therefore, practically, the Telecommunication 4.0 concept is the absolute consequence, with IT deepening into the CT industry in the current social industrial wave. For quite a long time, due to the fast development of the Internet industry, the communication industry fell into anxiety and there were doubts about whether the industry had a future. Under this circumstance, we should calm down and analyze the history, environment and general trend. The book, Telecommunication 4.0, illustrates the history of the communication technology development, explores the trends and summarizes the experience, which projects the prediction of the new systems and technologies. At the same time, the book also pays much attention to how to reverse the operation dilemma and how to reshape and form the industrial environment for the telecommunication industry. The book also highlights what is being popularized and what is happening, such as SDN and NFV, as well as the predictions for the industrial environment and commercial wave.

Telecommunication 4.0 created a new era for the development of communications networks and also offered an important opportunity for China to reshape the network system and international criterion segment. The orderly operation of human society depends on a series of rules and norms. Those who own the authority to set rules boast the capacity to improve the progress of society. Scientific research strength is the foundation to set scientific and technological criteria. However, in the past several technological revolutions during the world's modern history, China always acted as a learner and western science and technology were all along the most significant innovative origins. Therefore, western scholars have all along occupied the dominant positions in setting up scientific and technological criteria. The communication industry also faces this kind of embarrassing situation. China has more Internet users and mobile phone users than the rest of the world. But China applies communication criteria which are formulated mainly by European and American manufacturers. In order to change this situation, the Chinese communication industry strives to make headway. In recent years, in the new revolution of communication technology, China has been narrowing the gap between itself and the most advanced levels in the world. Especially, the fourth generation of mobile communication technology (4G) of TD-LTE independently developed by China has become international criteria, marking that China has reached advanced level in the world in the field of wireless communication technology.

Compared with the success we achieved in wireless communication system, China is still learning from other countries in core technologies, though the network system we own is the most advanced and the largest in scale, with the consequence that we hardly have any words in the network-system criteria. The international criterion, no matter how advanced it is, after all, cannot meet the need of the Chinese telecommunication network. If China builds its telecommunication network in accordance with western criteria, it will certainly bring about more operation costs and a large amount of customized requirements, a situation which would embarrass the Chinese telecommunication industry all the way. Therefore, for a long time, Chinese scholars have made painstaking efforts to research and explore, resulting in an abundance of inspiring works and experience. However, the future direction of network system development is hard to predict due to the complexity of the network system as well as constant changes of technologies and commercial development. Therefore, it will never be outdated to discuss this issue. The book, Telecommunication 4.0 deeply analyzes current condition of the telecommunication industry, with a purpose to figure out the direction of research and exploration of network technology. All of this will make great contributions to establishing the Chinese network criterion system.

Telecommunications 4.0 is already on the way, and the Chinese telecommunication industry is striding forward to the future.

Beijing, China   Houlin Zhao
                                                          Secretary-general of the International
                                                            Telecommunication Union (ITU)

Contents

Abbreviations

3GPP	3rd Generation Partnership Project
AC	Access Controller
ADSL	Asymmetric Digital Subscriber Line
AIC	AT&T Integrated Cloud
AMP	Advanced Manufacturing Partnership
AMPS	Advanced Mobile Phone Service
API	Application Programming Interface
AS	Application Server
ATCA	Advanced Telecom Computing Architecture
ATM	Asynchronous Transfer Mode
AWS	Amazon Web Services
B2C	Business to Customer
BGP	Border Gateway Protocol
BRAS	Broadband Remote Access Server
BYOD	Bring Your Own Device
C2C	Customer to Customer
CDMA	Code Division Multiple Access
CDN	Content Delivery Network
CI	Continuous Integration
CMNet	China Mobile Network
CNNIC	China Internet Network Information Center
CORD	Central Office Re-architected as Data Center
COTS	Commercial off-the-shelf
CPCI	Compact Peripheral Component Interconnect
CPS	Cyber-Physical System
C-RAN	Cloud of Radio Access Network
CRM	Customer Relationship Management
CS	Circuit Switch
CSCF	Call State Control Function
CT	Communication Technology

CTIA	Cellular Telecommunications Industry Association
DevOps	Development and Operation
DPDK	Data Plane Development Kit
DSN	Distributed Service Network
DSP	Digital Signal Processor
EPC	Evolved Packet Core
ERP	Enterprise Resource Planning
ETSI	European Telecommunication Standards Institute
ETSI NFV ISG	European Telecommunication Standards Institute Network Function Industry Specification Group
FDMA	Frequency Division Multiple Access
FPGA	Field-Programmable Gate Array
GBP	Group-Based Policy
GGSN	Gateway GPRS Support Node
GRE	Generic Routing Encapsulation
GSM	Global System for Mobile Communication
GSMA	Global System for Mobile Communications Alliance
HPE	Hewlett Packard Enterprise
HSS	Home Subscriber Server
HTTP	Hypertext transfer protocol
IA	Intel Architecture
IaaS	Infrastructure as a Service
ICT	Information and Communications Technology
IETF	Internet Engineering Task Force
IGP	Interior Gateway Protocol
IIC	The Industrial Internet Consortium
IMS	IP Multimedia Subsystem
IP	Internet Protocol
IPSEC	IP Security
IT	Information Technology
KPI	Key Performance Indicator
KVM	Kernel-based Virtual Machine
LTE	Long-Term Evolution
M2M	Machine-to-Machine
MANO	NFV Management and Orchestration
MES	Manufacturing Execute System
MME	Mobility Management Entity
MPLS	Multi-Protocol Label Switch
MSC	Mobile Switch Center
MVNO	Mobile Virtual Network Operator
NaaS	Network as a Service
NASA	National Aeronautics and Space Administration
NCP	Network Core Protocol
NFV	Network Function Virtualization
NFVI	NFV infrastructure

NMT	Nordic Mobile Telephone
OA	Office Automation
ODL	Open Daylight
OF-Config	Openflow Management and Configuration Protocol
OLT	Optical Line Terminal
ONF	Open Networking Foundation
ONOS	Open Network Operating System
OPNFV	Open Platform for NFV
OSPF	Open Shortest Path First
OSS/BSS	Operation Supporting System/Business Supporting System
OT	Operation Technology
OTT	Over the Top
OVS	Open vSwitch
OVSDB	Open Vswitch Database
PCEP	Path Computation Element Protocol
PCI	Peripheral Component Interconnect
PCIe	PCI Express
PCM	Pulse Code Modulation
PDH	Plesiochronous Digital Hierarchy
PICMG	PCI Industrial Computer Manufacturer's Group
POC	Proof of Concept
PON	Passive Optical Network
PPPoE	Point-to-Point Protocol over Ethernet
PS	Packet Switch
PTN	Packet Transport Network
QoS	Quality of Service
RCS	Rich Communication Suite
SAE-GW	System Architecture Evolution—Gateway
SBC	Session Border Controller
SDH	Synchronous Digital Hierarchy
SDN	Software Defined Network
SGSN	Serving GPRS Support Node
SoC	System on Chip
SONET	Synchronous Optical Network
SR	Service router
SR-IOV	Single Root I/O Virtualization
SSL	Secure Sockets Layer
STM	Synchronous Transfer Module
TACS	Total Access Communications System
TCAM	Ternary content addressable memory
TCP/IP	Transmission Control Protocol/Internet Protocol
TD-LTE	Time Division Long Term Evolution
TDM	Time Division Multiplexing
TIC	Telecommunications Integrated Could
TTM	Time to Market

V2X	Vehicles to X (Vehicles, person, infrastructure)
vCPE	Virtual Customer Premises Equipment
VDSL	Very High Speed Digital Subscriber Line
vEPC	Virtual Evolved Packet Core
vFW	Virtual Firewall
VIM	Virtual Infrastructure Management
vIMS	Virtual IP Multimedia Subsystem
VLAN	Visual LAN
vLB	Virtual Load Balance
VM	Virtual Machine
VNF	Virtual Network Function
VNFC	VNF Component
VNFM	VNF Manager
VoIP	Voice over Internet Protocol
VoLTE	Voice over LTE
VPC	Virtual Private Cloud
VPN	Virtual Private Network
VRF	Virtual Routing Forwarding
vSSL/IPSec GW	Virtual SSL/IPsec GW
VxLAN	Virtual eXtensible LAN
WDM	Wavelength Division Multiplexing
XMPP	Extensible Messaging and Presence Protocol

Chapter 1
Past and Present

Communication is one of the most basic demands for human beings. There has been a long history since human started to communicate with each other by different approaches. For thousands of years, people have passed information using figures, symbols, drums and fireworks. Further evolution includes speaking languages and those written on bamboo slip and paper. The non-stop development of communication keeps changing people's lives.

Dating back to the Chinese Shang Dynasty, there was a record of using beacon tower and drum for visual and audio communications respectively. In ancient Africa, beating drum is also the earliest means of delivery messages on record. By using a special kind of drum made of round timber, messages could be delivered with a distance of more than 50 km by relaying every 3–4 km. The messages were coded into specific patterns of beats. People have never stopped to explore new ways for communication by which messages can be delivered further and faster. Examples include carrier pigeon, beacon towers and post horse. There is no doubt that the time-effectiveness of information is very important. For instance, the legendary story of how Rothschild family made use of the information in the Battle of Waterloo has justified the importance of communication by itself.

In ancient times, the fastest way of delivering messages people could think of was using carrier pigeon, post horse and man power. As we all know, Marathon is in the memory of the Greek messenger who was deadly tired after running for 42 km. The Greek messenger, Phedippides, who went down in human history, always reminds us the hardship people have come across in the journey of pursuing communication technologies. Our ancestors believed that the speed for delivering messages is limited by means of physical movement and they would not be able to imagine how people live in modern society. This bottleneck was not unblocked until electric signal was discovered.

The emergence of electric signal brought dramatic change to the human society. In a sense, electric signal, as a medium for information transmission, replaced the formerly required physical movements. This opened up a new chapter of modern telecommunication. The evolution of modern telecommunication technologies is

© Springer Nature Singapore Pte Ltd. 2018
Z. Li, *Telecommunication 4.0*, DOI 10.1007/978-981-10-6301-5_1

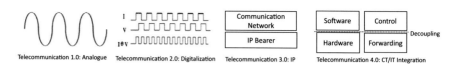

Telecommunication 1.0: Analogue   Telecommunication 2.0: Digitalization   Telecommunication 3.0: IP   Telecommunication 4.0: CT/IT Integration

Fig. 1 The development of key telecommunication technology

fascinating and several milestones have indeed made revolutionary breakthroughs. This spectacular process and technology development can be roughly divided into three phases, including the era of telecommunication 1.0, 2.0 and 3.0. Without the continuous accumulation and evolution, there is no chance for us to welcome the era of telecommunication 4.0. Consequently, it is necessary for us to look back and understand how telecommunication technology evolved during this magnificent process of innovation as shown in Fig. 1.

1 Pre-telecommunication 4.0 Era

– Telecommunication 1.0: Analogue Communication—The Origin of Modern Communication Technology

The field of communication technologies experienced a revolutionary change since the mid 19th century, thanks to the discovery of electromagnetic waves and invention of the telephone. In human history, information was transmitted by metallic wires for the first time, and electro-magnetic waves even enabled the propagation of information over open space. Since then, delivering messages no longer relied on visual and audio contacts. Electric signal, as a new type of carrier, boosted a series of technology innovation and opened up a new age of telecommunication in human history.

Figure 2 illustrates a schematic diagram of an analogue signal of the Telecommunication 1.0 Era. Analogue signal featured the key characteristic of this stage. The original signal was represented by using the amplitude, frequency and phase of sinusoidal waves, or the magnitude, width and sequences of impulses. Analogue signals are very common in daily life (i.e. voice signal, interference

Fig. 2 Analogue signals diagram

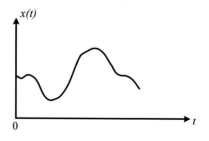

signal, noise and electric signal generated by teletube.). They all have a common feature of varied amplitude over time.

Analogue communication system is a process of generating analogue signals, transmitting them via electric signals, and recovering the signals at the receiving end. As shown in Fig. 3, the source generates voice signal (mechanical), which enters the modulator (such as telephone transmitters and photo-electric cells). The modulator further modulates the voice signal to continuous electric signal. One of the most popular modulation schemes is amplitude modulation (AM), where the signal is modulated using the amplitude of a sinusoidal wave. As shown in graph 4, the dashed curve illustrates the original voice signal whilst the solid one represents the modulated sinusoidal wave. The modulated electric signal may suffer from different levels of noises during its transmission over a certain medium. At the receiving end, the demodulator demodulates the signal according to certain parameters, restores the voice signal and finally hands it over to the destination. The modulated signal has varying amplitude and frequency according to the modulation scheme applied and characteristics of the original voice signal.

As the primary stage of communication technology development, analogue communication went through the following historic moments.

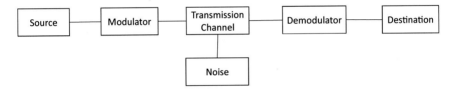

Fig. 3 Analogue transmission system

Fig. 4 Modulation signals
(amplitude modulation)

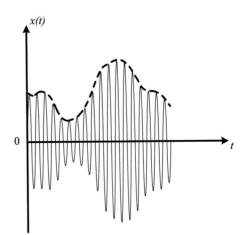

- In 1864, British physicist J.C. Maxwell established a set of electromagnetic theory and predicted the existence of electromagnetic waves. The theory indicates that electromagnetic wave and light have common characteristics and they both travel at the speed of light.
- In 1875, A.G. Bell, Scottish youngster, invented the first telephone in history. He patented his invention in 1876. In 1878, he succeeded in making the first long-distance phonecall between Boston and New York at a distance of 300 km. The well-known Bell Telephone Company was established soon after this successful experiment. Bell Telephone Company had an initial number of 778 customers, which increased to 3000 customers at the end of that year and 100,000 in 3 years. Twenty years later, the number of subscribed telephone in Bell's network reached 1.4 millions.
- In 1879, the first dedicated telephone switching system of went into operation. This established the foundation of switching technology. Almon B. Strowger, an undertaker in Kansas City, U.S., devoted himself to the investigation of an automatic switching device and succeeded. Almon patented his invention of automatic telephone exchange (Strowger switch) in March, 1891.
- In Nov. 1892, Strowger switch was deployed in La Porte, Indiana, United States. Hence, the first telephone office in the world was set up. Telecommunication stepped into a new era.
- In the field of radio communication, in 1888, German young physicist H.R. Hertz proved the existence of electromagnetic wave through a series of experiment using electric wave rings. His experiments also proved Maxwell's electromagnetic theory. Hertz's finding took the scientific community by storm. It is regarded as a significant milestone in the history of modern science and technology, which results in the birth of radio wave and greatly stimulated the progress of electronic technology.

The first generation of analogue cellular system was widely implemented for commercial use in 1980s, providing mostly analogue voice service. It used analogue modulation and Frequency Domain Multiple Access (FDMA) technology. The data rate is approximately 2.4 Kbps. Several successful systems include AMPS in North America, TACS in England and NMT in Northern Europe and etc.

1 Kbps = Kilobit per second − editor's note

Analogue technology was limited by its intrinsic disadvantages while bringing a roaring success to communication technology. For instance, it is vulnerable to interferences so that the mixed signal is hard to be restored. This may result in poor phone call quality. Moreover, phone call could be easily hacked since analogue modulation is simple and non-encrypted. The content of a phone call can be easily seized by simply make a copy of the analogue signal. In addition, the development of telecommunication 1.0 technology was also restricted subject to the limited network capacity, incapability of providing data service, low spectrum efficiency and high costs, etc.

– Telecommunication 2.0: Digital Communication—The Myth of Shannon

In order to overcome the inherent defects in analogue system, digital communication technology emerged. This also marked arrival of the era of Telecommunication 2.0, in which digital communication played a key role.

Complex information is represented by simple binary numbers in digital communication. The beacon-tower communication in ancient times could be regarded as a simple form of digital communication. The on and off statuses of the beacon towers represented "1" and "0", indicating whether there was an invasion. The Morse code which is still in use today encodes text information to various combinations of dots and dashes, which can also be considered as a variant form of binary "1" and "0".

Telecommunication 2.0 referred in this book indicates modern digital communication. This was remarked by the invention of PCM (pulse-code modulation) in 1937, which laid the foundation of modern digital communication. As the PCM technology continued to evolve and development, digital communication has become the core of communication technology which was still been widely used today.

Digital signal is derived from analogue signal by sampling, quantization and coding. The digitization of analogue signal realizes the transition from continuous waveform to discrete data samples, which are normally quantized sampling time and wave amplitude. An example of digitalizing is given in Fig. 5. First of all, signal is no longer continuous in the time domain. Samples are taken by using impulses with a certain time interval. As shown in Fig. 5, sampling rate (the reverse of sampling time interval) is a key parameter in this process. Naturally, a higher sampling rate gives a smaller sampling interval. As a consequence, the digitized signal is more accurate with higher quality. Secondly, the samples also have discrete amplitudes, which are represented by binary codes with certain quantization

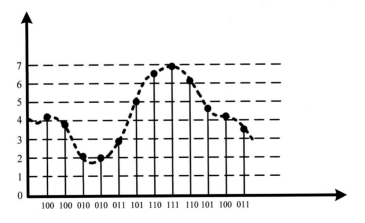

Fig. 5 The digitalization of analogue signals

accuracy. Different representations make up different coding schemes. As an example, a 3-bit binary coding scheme is used in Fig. 5. This scheme provides a maximum of 8 levels of amplitude with 8 different 3-bit binary numbers. Reasonable quantization algorithm should be used for particular cases to minimize the accuracy. Thus, the original signal can be restored at the receiver end with corresponding decoding algorithm.

Digital communication will not be possible without the pioneering research work down by Claude Shannon. Known as the "the father of information theory", Shannon has made great contributions to the development of communication technologies. He is most famous for having founded information theory. Shannon's theory quantizes the information, of which the definition is considered vague for hundreds of years. It makes the information measureable and defines the maximum capacity of a given transmission channel. This is a remarkable achievement in the history of communication technologies. The Shannon Capacity equation is shown as below,

$$C = B\log\left(1 + \frac{S}{\sigma^2}\right)$$

where S is the average received signal power, measured in watts (or volts squared). B is the bandwidth of the channel in hertz and σ^2 is the average power of the noise.

Thus it can be seen that the maximum rate at which information can be transmitted over a given communication channel is dependent on the specified bandwidth and signal-to-noise ratio. The channel capacity is a tight upper bound reference to evaluate if a certain communication technology is optimized. Referring to this upper bound, people have been working towards the limit by making the best of the channel resources.

During the process of digital communication development, digital mobile communication has experienced a rapid growth and gradually developed into a rather independent field. The most well-know technologies include GSM (Global System for Mobile Communications) and CDMA (Code Division Multiple Access) developed by Europe and the U.S. respectively. Compared to analog communication, digital communication overcomes some of its weaknesses. First of all, digital signal has strong interference-resistance property and does not suffer from noise accumulation. The interference caused by noise can be overcome by exploiting specific modulation and coding schemes and corresponding decision mechanism. Besides, the encryption of digital signals is easier and more flexible, which provides more reliable transmission. As the hardware techniques including integrated circuit continue to evolve and upgrade, the size of the communication equipment in the era of digital communication is much smaller. Meanwhile the costs are also remarkably reduced.

There are numerous milestones that worth mentioning during the development of digital communication in human history.

- In 1837, Morse invented telegraph that could transmit text information with a series of dots and lines, which was regarded as the earliest form digital signal.
- In 1924, Harry Nyquist proposed the sampling theorem, which defines the relation between the sampling frequency and signal bandwidth. It established the theoretical foundation for the digitization of continuous-time signal in the form of discrete samples.
- In 1937, PCM (Pulse Code Modulation) is invented, which enables the transform of the signal from analogue to digital. This fundamental technology laid the foundation for the modern digital communications.
- In 1947, Bell Laboratory successfully developed a modulator with 24 electronic valves for experimental use, which demonstrated the feasibility of PCM.
- In the 1950s, the digital microwave communication technology started to emerge. The first generation of low frequency and low bandwidth digital microwave system was developed in the early 1970s. Many countries including the U.S., Australia, Canada, France, Italy and Japan deployed the 4 GHz digital microwave relay system in their backbone networks.
- In the 1980s, the second generation of digital microwave system was put into commercialization. It is worth to note that both first and the second generations of digital microwave systems reside in the domain of PDH (Plesiochronous Digital Hierarchy). PDH suffers from complex transmission processes, weak OAM and low efficiency, which had certain limitation in practical applications.
- In the mid-1980s, the birth of SDH (Synchronous Digital Hierarchy) marked the start of third generation of digital microwave system. In 1988, the SDH standard was proposed on the basis of the SONET (Synchronous Optical Networking) by ITU (International Telecommunication Union). Compared to the PDH, SDH consolidated the standards in Europe and North America, allowing global interoperations at the rate of STM-1 or higher.
- The SDH features higher spectrum efficiency, transmission quality, reliability and accessibility. It is more advanced in many aspects including transmission rate, frame structure, signal multiplexing and de-multiplexing, frequency allocation, interfaces and network management. SDH can form a loopback with fiber link, or act as the protection channel of the fiber link. It is also feasible to directly implement SDH links and networks, etc. The SDH microwave system has experienced a rapid growth in 1990s.
- In the mid-1980s, second generation mobile systems (2G) emerged. The most well-known standard are GSM and CDMA from Europe and the U.S. respectively, both of with were widely used in many countries. This marked the maturity of wireless digital communication technologies.

GSM was originated in the Europe. As early as the 1970s, some developed western countries have started to develop digital mobile communication system. In 1991, the GSM 900 MHz cellular system was first deployed in Europe. Since then, mobile communication stepped into the era of telecommunication 2.0. With the development of new devices and the establishment of cellular networks, GSM stood out for its uniform interface standards and well-defined protocols. In 1988,

Qualcomm proposed CDMA technology widely inspiring the community. CDMA is another variation of digital communication, and was based on spread spectrum communication. In terms of technical advancement, GSM and CDMA all have its own pros and cons. GSM is more mature due to its early and wide application and rich experience on operation, while CDMA is more efficient and outperforms GSM in security issues. CDMA had more competitive edges in terms of techniques, but the roaming issues have set boundaries for its global implementation.

At the same time, China has witnessed a great improvement in the revolution of communication technologies.

- In 1979, China constructed the first PDH microwave link (Beijing-Wuhan).
- In 1989, China constructed 6 GHz 140 Mbps[1] PDH microwave link between Beijing and Shanghai.
- In 1992, the first GSM cellular communication system in China came into operation in Jiaxing, Zhejiang.
- On Sep 19, 1993, the first digital mobile communication network in China was setup and under operation in Jiaxing.

As the market became mature with increasing number of subscribers, the existing network technology gradually became insufficient. The traditional network based on TDM has a low capacity, thus no longer support growing number of users. The network designed only for audio transmission was also incapable to support varied kinds of services that started to emerge. Thus a new type of network is required, which has to guarantee the quality of service while support more users and services with simple network structure and lower cost. The strong demand for this new network triggered another revolution in communication technology.

– Telecommunication 3.0: IP-based Communication Network—The One had the Last Laugh

In order to solve problems in traditional TDM systems, the communication network is in urgent need of novel technologies. IP-based communication network was not a preset goal, but a technical decision after countless debates and practices. We have to look back to the mid-20th century in order to understand how IP eventually stood out.

(1) IP wins in the competition with the ATM (Asynchronous Transfer Mode)

The full name of IP is Internet Protocol, whose appearance is closely related with the internet. The predecessor of the internet is the trial network APRANET of the US military, which was built up in 1969 to realize a distributed interconnecting system with non-centralized control. APRANET adopted the NCP (Network Control Protocol) in its early times, which was the early prototype of the TCP/IP (Transmission Control Protocol/Internet Protocol). In 1974, TCP/IP was officially launched. In 1981, IPv4 (4th version of the Internet Protocol) was published in the standard RFC (Request for Comments) 791. Later in 1983, the entire APRANET

---

[1] 1 Mbps means million bits per second.

shifted to TCP/IP, and gradually evolved to the global internet. TCP/IP thus continued to be used, establishing the basis of internet communication.

Since the 1990s, the internet has witnessed dramatic development around the globe, starting from the simple e-mail to all kinds of instant messaging, from online games to virtual communities and from communication services to internet finance, penetrating into all aspects of human lives within 30 years. It has become an indispensible part of the modern society. Global operators adopted the IP technology to build up many dedicated networks. For example, China Mobile built "dedicated IP network" and "CMNet (China Mobile Network)". China Mobile's IP network, as part of the global network, also develops rapidly. The CMNet, which targeted at basic internet services and the "dedicated IP network", which targeted at high-value services gradually become a large-scale IP-based network covering the whole nation as well as some popular locations around the world.

Another technology coexisted with IP at almost the same time was ATM. And there was the famous "Debates over IP and ATM" aimed at unifying the basis of the communication network. The initial standard of ATM was proposed by the ITU-T, with many typical characteristics of telecommunication network protocols, aiming at connection oriented service, high reliability and QoS. These raised tradeoffs such as protocol complexity and high costs, making it hard to fit in the explosive growth of bandwidth and scale of the internet services. Fundamentally, this technology argument was a collision between different methodology and thinking, respectively Internet-wise and telecommunication-wise. Although IP won the debate, these two technologies are actually mutual-complimentary in many ways. As IP technology continued to develop, it borrowed some ideas from the ATM, such as the MPLS (Multi-Protocol Label Switching) and QoS (Quality of Service).

The emergence of internet services such as VoIP (Voice over Internet Protocol), IPTV (Internet Protocol Television), remote education, mobile health and e-commerce indicates the arrival of a new era of "everything over IP". As the IP technology stands out, the telecommunication network is also gradually evolving to an IP-based system. This evolution is illustrated in Fig. 6.

(2) **Towards a comprehensively IP-based telecommunication network**

The IP-based telecommunication network system service includes two major parts, namely switching system and accessing system. Both systems further include fixed and mobile subsystems.

(a) **IP-based switching system**

Switches are key elements in the telecommunication network, providing many core functions in the forwarding plane. The switching system has been playing an essential role in the telecommunication network since its birth. The history of the switching system can be divided into three stages, i.e. manual switching, electromechanical switching and electronic switching. Thanks to the emergence of semiconductor and computer technologies, the switching system gradually evolved from manual toward electronic, programmatic and digitized control. Since mobile network came much later than the fix network, it switching system adopts

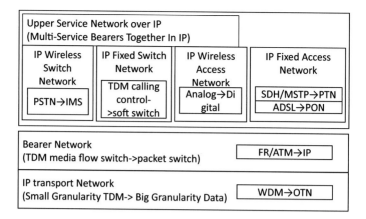

Fig. 6 The evolvement of the IP-based communication network

programmatic and digitized technology since deployment. However, the capacity of the switching matrix and the number ports are limited under the TDM switching regime. Hence, the switching mode becomes the bottleneck of the overall capacity.

As the transition to IP-based softswitch system continued to take place, VoIP service was realized in mobile communication network. The separation of the control plane and forwarding plane makes it easy for operators to conduct centralized deployments and management whilst increasing the overall connectivity and capacity of the switching system. Naturally, the IP-based devices became the develop trend of the modern switches. As the technology develops, the successful introduction of IMS (IP Multimedia Subsystem) marks the thorough realization of the IP-based core network. IMS is similar to softswitch in many ways. First, the two of them are all based on the packet network, and have all realized the separation of control and forwarding plans. However, softswitch is deployed in the CS (Circuit Switching) domain, whereas IMS is on in the PS (Packet Switching) domain. IMS enables the PS to share some functions of the CS, and supports multimedia services whether or not based on session, hence IMS is naturally intrinsically an IP-based entity. Benefited from the wide deployment of IMS, the core network has transformed to a system completely based on IP. A further end-to-end IP transformation is under the way.

(b) **IP-based access network**

As the transition of switching system to IP is more or less completed, it's time to consider the end-to-end IP-based network. This is the most difficult and essential part of the overall telecommunication system evolution.

The mobile access network was designed to be IP-based in backhaul network since 3G (Third Generation Mobile Communication Technology). As the access network was completely IP-based since then, the entire end-to-end telecommunication network was based on IP. Later on, network elements were designed with IP capabilities from the beginning. Hence the network, no matter core or access, was

IP-based. There is no need to worry about the compatibility anymore. Multimedia services whether or not based on session can be provided within the PS domain. Hence the entire telecommunication network transformed to IP technology.

Meanwhile, the fixed access network has developed from ADSL (Asymmetrical Digital Subscriber Loop) to VDSL (Very High Bit rate Digital Subscriber Loop) and all the way to PON (Passive Optical Network). This is also a process of adapting to higher bandwidth and IP technology. For the coming FTTx era, IP + optics has become one of the main stream technologies.

(3) **IP evolution of the transport network**

As the IP-based network constantly expands and the internet service develops, it is important to investigate how transport network also adapts to IP technologies.

Before IP, SDH dominated the transport network. The TDM technology suffered from high deployment cost, low resource and transmission efficiency. It gradually became the bottleneck of the transport network. To overcome this, operators like China Mobile has to make a choice for the IP technology which is suitable for its network architecture.

However, the best effort IP technology treats each of the packets equally. It is lack of necessary guarantees for transmission performance including latency, jitter, packet error rate and bit error rate. Therefore, transport network must seek new means to keep its carrier-grade OAM in the process of embracing IP technology. For instance, as for backbone network, it evolved from the earlier IP over SDH over WDM (Wavelength Division Multiplexing) to IP over WDM by directly modulating the IP data to the WDM. This approach reassigned the SDH-layer switching functionality to the IP layer with WDM purely providing the physical channel.

In order to guarantee the reliability, OTN (optical transport network) is introduced for its robust protection and management mechanisms. This realized another evolution from IP over WDM to IP over OTN.

As for metropolitan layer, SDH technology based on rigid channels was not sufficient for new service, high bandwidth and more complex and burst traffic flows. Therefore, the SDH was gradually replaced by the PTN (Packet Transport Network) technology, with the characteristics of both the IP/MPLS forwarding, robust transmission reliability and OAM.

As shown in Fig. 6, The IP-based telecommunication system went through a series of milestones. The trend shown in Fig. 7 takes the evolution of voice service over IP based network in China Mobile as an example.

- In 2006, softswitch in mobile network started to adapt IP technology. The links interconnecting the servers of mobile switching centre and media gateway gradually migrate to IP-based technologies.
- In 2009, the 2G (Second Generation Mobile Communication Technology) mobile access network began its IP adoption, i.e. making the interface between the server and base station controller IP-enabled.

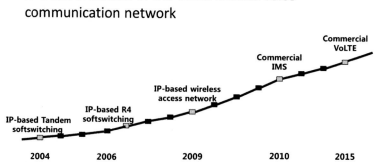

Fig. 7 History of voice service over IP-based network in China mobile

- In 2010, an IMS system delicately designed according to IP officially went into commercialization. Fixed/mobile-convergence core network thoroughly adopted IP technology
- In 2015, VoLTE (LTE-carried voice) is launched for commercialization. This is yet another milestone, which marks the completion of IP adoption of the end-to-end telecommunication network.

Simple, practical and expandable are the main features of IP technology. It has enabled the decoupling of top-level application and bottom-layer physical networks. The worldwide adoption of IP technology sharply reduced the costs of telecommunication equipment and enhanced the telecommunication network with the advantage of maturity and high bandwidth. As technology develops faster and faster, telecommunication network has entered the Post-Moore's Law period. Rapid evolution of technologies accelerated the upgrades of the telecommunication equipment. The explosively increasing number of subscribers also poses a greater challenge of extremely high flexibility and agile deployment.

As for operators, the massive amount of network resources may become a heavy burden if they cannot construct smart, elastic, self-organized and self-recoverable networks.

On one hand, the explosive increase of data traffic has exceeded the Moore Law's, going up at an exponential rate each year. On the other hand, the requirements for customized and individualized service is also growing, establishing a fundamental challenge for the rapid response of the network configuration.

Though IP technologies meet the demands of basic network function and performance, a low cost, high flexibility, smart network with elastic capacity and fast deployment ability is still preferred. Reduced costs and high flexibility makes it easier for the network to be upgraded, optimized and deployed, whilst elastic capacity and fast deployment enable dynamic resource allocation to tackle with intensively growing traffic. These all call for the emergence of a new generation of telecommunication network, indicating the coming of the 4.0 era.

2 Entering the Telecommunication 4.0 Era

For long, demands and technology have been like two cars racing on the motorway, taking the leading position back and forth. For most of the time, demands seem to be ahead of the technologies. When the gap between demand and technology is too wide, it is usually not far from a technology explosion, with which technology are triggered to overtake demand again. In fact, we are standing in the process right now, on the eve of Telecommunication 4.0.

The Telecommunication 4.0 era features the following key characteristics and development requirements.

1. Agility is the intrinsic characteristic of Telecommunication 4.0

Before Telecommunication 4.0, telecommunication networks were composed of individual equipments with integrated software and hardware. The connections between these equipments are extremely complicated. The functions of each equipment are dedicatedly designed for specific use cases and are normally static. Therefore, operators need to upgrade or replace the legacy equipments in order to introduce new features. This is an extremely costly, complex, tedious and time-consuming process, including the preparation of the machine room, installation and testing. This has caused delayed response of operators to the new technologies, which does not satisfy the requirement of customized services. When customers subscribe to a new service, they always want it to be deployed immediately. This is hard for operators to couple with since the large-scale traditional network does not react efficiently for the deployment of new services.

To some extent, changing the network to adapt new requirement is not a choice anymore. We need a vision for the future and Telecommunication 4.0 is a desired solution.

In the era of Telecommunication 4.0, the core feature of the operators' network is the ability of realize series of network function using software installed on the general-purpose hardware. The potential upgrades can be done simply by updating the software. The hardware does not need to be replaced or upgrade. This avoids uncontrollable expansion of network scale and it greatly benefits the agile deployment of network functions. Hence, telecommunication 4.0 is the era of agile network.

2. Openness is the key feature of Telecommunication 4.0

Before Telecommunication 4.0, the network was a relatively closed system, where we had to provide various platforms and dedicated protocols for different services. The new technologies, diverse network functions and increasing requirements from users have urged operators to continuously enhance the scale and performance of the network. As a result, the network undoubtedly became more and more complicated, including both bottom layer physical platform and or top layer routing and communication protocols. To solve this problem, a network with a simplified structure, strong capability and elastic capacity is in great need. The coming of the

Telecommunication 4.0 era exactly solves this problem. Based on the two core technologies, namely NFV and SDN, it realizes the exposure of network capabilities through providing APIs (Application Program Interface). This greatly reduces the complexity for content provider to use the network resources of operators.

In Telecommunication 4.0, the decoupling between the software and hardware for communication equipment will be realized, making it possible for general-purpose IT hardware, virtual machine software and resource management software to be deployed in the operator's network. The entire communication industrial chain shifts from simplex chip and system vendors to a combination of chip, system, IT hardware and IT general software vendors. The former industry which was relatively isolated transforms to a co-operative and synthesized one. This restructured industrial chain is believed to boost the development of Telecommunication 4.0. Hence, Telecommunication 4.0 marks the openness of the telecommunication industry.

3. Software (virtualization) is the core implication 4.0 communication

Before Telecommunication 4.0, the operator's network was composed of large numbers of dedicated communication equipments including routers and switches. One important characteristic of such dedicated equipments is the tight attachment of the software and hardware. These equipments appear to be in different forms according to specific purposes and are deployed in distributed machine rooms. As new services are added to the network, new equipments need to be deployed to substitute the legacy ones. Meanwhile, network expansion normally indicates the purchase of a new set of equipment, or the replacement of equipment with smaller capacity to larger. The constant replacement of hardware requires large investment, including the cost for replacing equipment, labor for operation and maintenance, new machine room and electricity bills in one hand. It also causes operator's slow reaction to new services due to the long development period of hardware equipments. Besides, the replacement of equipments needs to be in line with the requirements and the later production of the hardware. It usually takes 2 or 3 years to fully develop a dedicated piece of hardware. Such long development period results in the extremely slow replacement rate of the traditional communication equipments, which causes delay to the response of operators to emerging new services. At the same time, it can be seen that internet industry based on software technology and service innovation is experiencing a booming development. Users' demand for new services increases. Internet enterprises' responding speed is also getting faster. The users' urgent needs for new services and operators' slow reaction in getting new functions online has become a key disappointment in the traditional communication network.

Stepping into the Telecommunication 4.0 era, the development of communication industry will demonstrate a trend of "softwarization". More and more elements in the communication network can be realized by software based on general-purpose hardware. Compared with the conventional system based on dedicated hardware, the software system transforms the lifecycle of the establishment of a new communication service. The traditional hardware development and integration is replaced by software engineering. This makes it possible for more general developers to be involved. As a consequence, it is also much faster for the system to be integrated and deployed. Virtualization of the network function can considerably shorten development period of a new communication service, accelerate the roll-out of service and improve the efficiency of service innovation.

4. The integration of the IT and CT industries is the essential vision of Telecommunication 4.0

The development of the communication technology is a process of learning, improving and innovating. Generally speaking, the development track of CT and IT are separated before Telecommunication 4.0 although advanced IT such as IP is introduced to CT. There were relatively few areas in which these two fields deeply interact.

"Integration" is the keyword of Telecommunication 4.0, and the integration of the IT and CT industries is the essential vision of Telecommunication 4.0.

First, there is the comprehensive integration of IT and CT. The core technology of Telecommunication 4.0 is SDN and NFV, which by themselves are the result of the comprehensive integration between IT and CT.

SDN originated from IT, and was firstly applied in data center and router network for the purpose of routing configuration. It was later used in telecommunication networks. NFV technology originated from cloud computing, and extends as the key technology for virtualization of the telecommunication network.

Second, the integration of technologies also drives the integration of industries, where CT and IT share a overlapped eco-system. Telecommunication 4.0 will introduce more IT stakeholders (such as virtual platform providers, hardware providers, cloud management platforms and SDN controller providers). As thorough competition and cooperation starts to take place, a brand-new Telecommunication 4.0 eco-system is going to be established.

Besides, as the industry integrates and people from different communities join together, exchanging ideas between IT and CT will also become a norm. Participants in Telecommunication 4.0 will have to study and understand the manners and rules of different industries. As a result, new mode of thinking and collaboration will be formed in Telecommunication 4.0.

Telecommunication 4.0 is an opportunity for the industry in the next decade. It is a "one in life time" opportunity for the participants to make a change in the industry. Telecommunication 4.0 has just sprouted. People should be aware of the trend, facing up to the challenges, opening minds for innovative ideas and embracing the evolution of telecommunication industry for another promising journey.

Chapter 2
Telecommunication 4.0 Development Pursuit and Vision

As social animals, Communication between human beings is thus the essential condition for the existence and development of human society and also a fundamental social demand for individual human beings.

In 1943, Abraham Maslow, the American humanistic psychologist, published <*A Theory of Human Motivation*>, which elaborated Maslow's needs model. According to the theory, once the low-level needs are met, high-level needs will naturally emerge. The theory can also be applied to the communication needs of human beings. The primitive communication system such as letters could no longer satisfy the need of instant communication nowadays. And then this problem was solved by the modern telephone and telegraphs system thus the communication needs of human beings were pushed to a higher level. After the cellular telephone solved the general communication needs between people, the basic communication systems can no longer satisfy man's communication needs. More abundant information, massive connection and smarter capability become the future directions.

1 The New Maslow's Model for Communication: Hierarchical Communications Requirements Determine the Telecommunications Development Phase

Inspired by Maslow's needs model, we are happy to find that human beings communication needs can also been seemed as a hierarchical model, which can be divided into 5 levels as shown in Fig. 1: necessary communication, common communication, information consumption, sense extension, and self-liberation. Same as Maslow's needs model, these communication needs should also be satisfied level by level. The higher the communication level is, the more dependent on inner realization. And the more unbounded the need is, the higher its added-value is.

© Springer Nature Singapore Pte Ltd. 2018
Z. Li, *Telecommunication 4.0*, DOI 10.1007/978-981-10-6301-5_2

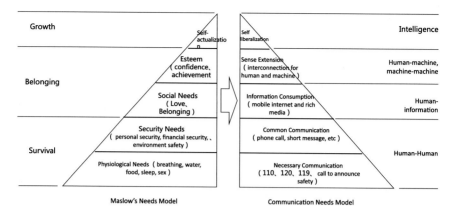

Fig. 1 Maslow's hierarchy of needs and the hierarchy of communication needs

- Necessary communication: focus on the necessary communication needs of human beings. Due to the shortage and the high cost of means of communication, this level only meets the most urgent and necessary communication needs.
- Common communication: focus on the general communication needs of human beings. With the decrease of costs and the increase of communication tools, people can easily connect with each other and the amount of communication traffic spreads at a rapid rate.
- Information consumption: focus on the communication needs between human beings and information. People are no longer satisfied with the connection between each other. The core need in this level is the consumption of information and the increasing variety of communication tools, which will lead to the sharp increase of needs for internet broadband.
- Sense extension: focus on human-to-machine and machine-to-machine communication needs. Based on the information consumption, man's sensory experience will further be extended and the communication network will cover all corner of the world. The key need of this level is the interconnection between human and machine. The number of network connections will rapidly increase.
- Self liberation: the previous four levels focus on the scale and type of connection. This level, however, pays more attention to the knowledge produced by the improvement of the connection. The communication system will become smart and thus enable human beings to realize self-liberation.

According to Maslow's theory, the needs will drive behaviors while behaviors in turn satisfy the needs. The needs and behavior form an ascending spiral circle. To the communication needs model, the communication needs and communication systems form a circular relationship, that is to say the emergence of needs pushes the development of communication technologies and communication systems, while the improvement of the communication system pushes the communication needs to a higher level. Corresponding to the hierarchy of communication needs,

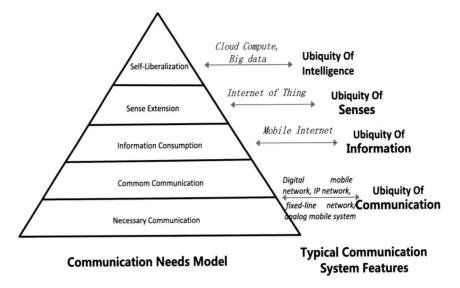

Fig. 2 Four levels of the communication system and their respective features

the typical human communication systems can be divided into four major phases, i.e., ubiquity of communication, ubiquity of information, ubiquity of senses, and ubiquity of intelligence, as shown in Fig. 2.

Currently, the growth of global population has slowed down. In developed countries and emerging market economics, which take the dominant roles in the global economy, their population has reached the peak and demographic turning point has coming. After years of development, the penetration rate of telephone and internet is becoming saturated. Meanwhile, the necessary communication need and the common communication need have been satisfied basically. The ubiquity of communication has thus been achieved. In the future, the communication need will move to a higher level and enter the information consumption era.

2   Ubiquity of Information Will Meet 3rd Level Need in the Model of Communication Needs

Once the fundamental communication needs among people have been satisfied, higher level ones will come out naturally. People needs for the acquisition and consumption of abundant information have become stronger as well as for the diversity and flexibility of communication methods.

With the rapid development of modern science and technology, human beings now have much more information and scientific knowledge. Since the 21st century, human beings have entered the information explosion era where all kinds of information, such as news, entertainment information, advertisement and scientific information have been greatly enriched. This information explosion has enriched our everyday lives on one hand, and inspires us to further explore the world and ourselves on the other hand. Consequently, the information storage needs of human beings keeps growing.

According to the statistics from a British scholar James Martin, the sum total of human knowledge by 1800 was doubling every 50 years, by 1950, doubling every 20 years, by 1970, doubling every 10 years, by 1980, doubling every 3 years. Throughout the world, as much as 13,000–14,000 essays are published every day, 700,000 new patents are registered every year, and more than 500,000 kinds of books are published every year. New theory, new material, new skills and new methods constantly appear and accelerate the ageing of knowledge. Statistics also shows that the half-life period of one's knowledge was 80–90 years in 18th century, 30 years during 19th and 20th century, 15 years in 1960s. But after 1980s, it was reduced to 5 years. Some reports indicated that the amount of information printed around the world doubles every 5 years. Nowadays, the information amount in the *New York Times* in one week is as much as the total information amount one scholar in 17th century could receive in his whole life. The information amount produced in the last 30 years is more than that of the past 5000 years together.

With the rapid growth of the amount of information, the propagation medium is enriched continually. In the last time, the medium mainly depends on books and newspapers. But thanks to the rapid development of electronic technologies, the medium is extended to Videos, digital books and digital newspapers. Meanwhile, with the development of internet and mobile internet, the speed and scale of information acquisition has reached to an unprecedented level. Now information can be shared and exchanged all throughout the world and the communication network has become an essential infrastructure in the information society.

The acceleration of information acquisition and spread has resulted in the growth and explosion of communication network traffic. According to the Ministry of Industry and Information Technology of China, the internet traffic from cell phones in China reached 432 million G-bits at the first two months of 2015. And the number has been doubling for 2 successive months was accounting for 89.3% of total mobile internet traffic. The GSMA's Mobile Economy Report, citing Cisco's Global Mobile Data Traffic Forecast 2015, claimed that during the period from 2015 to 2020, the compound annual growth rate of global mobile traffic is expected to be around 40–50%, and by 2020, the mobile traffic will be 9–10 times more than that of in 2014, as shown in Fig. 3.

All operators around the world are accelerating the deployment of communication networks in response to the traffic explosion. Since 2013, operators have started to extensively build 4G LTE networks. Taking China Mobile as an example, from the end of 2013 to the end of 2015, China Mobile built over 1 million 4G base stations, creating the largest 4G network around the world which can support more than 1 billion people. Such development of the 4G network greatly extends the channels for information consumption.

Fig. 3 Global mobile data traffic forecast

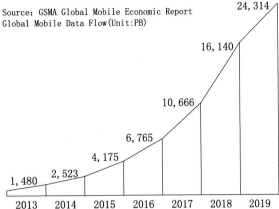

Source: GSMA Global Mobile Economic Report
Global Mobile Data Flow(Unit:PB)

The traditional simple and extensive development mode can no longer meet the needs of the explosive growth of traffic any more. Only the communication network with Information Ubiquity feature can satisfy the need at this stage. This needs the network can carry large amounts of information and support flexible network schedule capability.

On one hand, it requires the network to develop in a low-cost and high-efficient way. Only the new technology and new architecture with lowest cost and greater bearer capacity can meet the needs of great growth of the net traffic.

On the other hand, traditional operators can hardly handle the need of agile network scheduling requirement resulting from the traffic fluctuation and the changeable traffic flows. Thus, the future communication network should be capable of supporting the exact distribution of traffic. And by the refined dynamic schedule, the future network should be a resource-saving network.

3 Sense Extension Will Meet 4th Level Need in the Model of Communication Needs

Information consumption allows human beings to better understand the world and themselves, narrows the distance between man and man, man and animals, as well as man and the world. And it makes the information and knowledge is close enough to touch for everyone. Nevertheless, human beings will never stop exploring and conquering the world and are not satisfied with the known knowledge and the understanding and control of the known world. Human beings are always expecting to expand their sense to each corner of the world, and even to all the sands in the world.

Objectively, many places in the world, such as the deep seabed, the high mountains, the faraway Arctic, and the vast universe, are out of reach of human beings or unsuitable for long-time stay due to the hostile environments. However

Smart equipment and sensors are able to keep working under extreme conditions. Just like our fingers, they can help us to know and monitor the environment, and also provide real-time environmental reference data. All smart equipment and sensors can thus be regarded as an extension of the human body.

When human being's sensory capability covers every corner of the physical world, the spreading and sharing of information cannot be achieved without the interconnectivity of related equipment through the network. Only with the full interconnectivity between man and machine and between machine and machine, the variety and timeliness of the information for human beings can be promoted greatly as shown in Fig. 4.

The Internet of Things complies with the trend of Internet of Everything. And it connects everything with relevant equipment in one region into a global network to realize the recognition, position, management, and control of each node in the network. Nowadays, the Internet of Things is mainly applied in data acquisition, mobile positioning, automatic control, and daily services. With the improvement of communication, sensory and control capability, the field of application will be further extended.

Currently, the Internet of Things has just started up. It is estimated that in 2020, 25 billion equipments will be connected to the network and the amount is equivalent to 4 times of the world population. With its development, the amount of

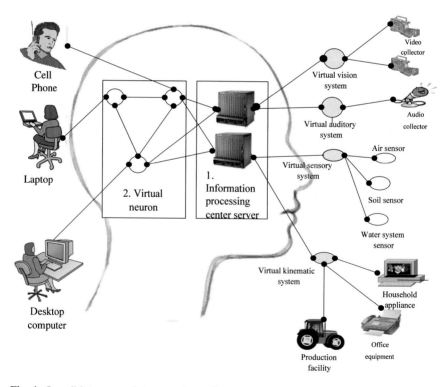

Fig. 4 Overall interconnectivity extends men's sensory network

Fig. 5 Growth of the APP numbers

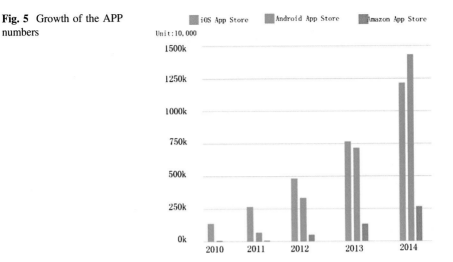

equipment connected will further increase. By 2025, the number is expected to reach 100 billion.

The extension of Information Ubiquity and Sense Ubiquity also promotes the development of smart phones and the smart APPs, as shown in Fig. 5. These creative APPs lower the costs of information acquisition and simplify the information acquisition mode, thus making it more convenient for human beings to control things. According to the statistics of the APP market from iiMedia Research, by the second quarter in 2015, the number of APPs for the iOS system has reached 1.21 million, for Android 1.43 million and for the Microsoft system over 0.3 million.

The need of Internet of Things raised higher requirements for the communication network, as demonstrated in the following four aspects:

- Massive connections will bring great impacts on the communication network. The huge number of connections needs to create substantial dynamic links and need the network to have great flexibility.
- Diverse connection types and transmission contents require the communication network to be differentiated and personalized. The internet of things, with a wide coverage, involves many types of equipment with different features. The communication network hence faces great challenges. For example, in the scenario of electronic meter reading for families, terminals of the internet of things only need to transmit a small amount of data regularly, and the requirement for connection is thus fairly low. In telemedicine scenario, however, several connections need to be created at the equipment, including audio and video connections, sensor connections, and instrument control connections and all the connections shall be highly synchronized. Those requirements are beyond the capability of the current communication network, and thus new fragile network architecture is needed.

- The connections among different nodes of the internet of things are quite complicated. There exist point-to-point connections, tree-structured connections, star-structured connections, and even net-structured connections. Meanwhile, the equipment of the internet of things in the tree, star or net-structured connections can also serve as nodes of higher-level networks and further form more complicated connections. The current single layer network control mechanism can hardly meet the need of the grid type network of the internet of things. A more open network architecture is needed.
- The connections should be low-cost and energy-efficient. The communication network needs not only to support the Internet of Things, but also to lower the cost of network equipment.

4 Ubiquity of Intelligence Will Meet 5th Level Needs in the Model of Communication Needs

In the stage of Sense Ubiquity, man's sense has already reached to the whole world and deepened man's understands of the world. But, human beings still have to make substantial intelligent efforts to study the world that is yet to be known actively and do active research on the objects and machines that need to be controlled. The objects and machines controlled by men cannot control themselves or learn by themselves, nor can they proactively communicate with human beings to help solving our problems. Human beings, while enjoying the benefit of the information era, have also been fall into the maelstrom cause by explosive amount of information. The world is filled by information garbage. To find out useful information and help human beings out of the information maze, artificial intelligence is the best way to help human beings to liberalize themselves and realize the ultimate pursuit of a intelligent world.

The need of self-liberalization is actually the need of making people "lazy", which mainly includes two major aspects:

(1) Intelligent needs: objects and networks will be endowed with intelligence through scientific approaches like artificial intelligence and neuron so that they can learn, feel, improve, and maintain themselves. Intelligent and labor work for men can thus be reduced. (2) Active needs: objects and networks will have the capability of providing active service through big data mining and human behavior tracking. Or they can provide more convenient and comfortable service without commands or only with few mutual communications.

The key technologies for the communication needs of the 5th level include artificial intelligence and neural networks and the latter can be seemed as a branch of the former. Artificial intelligence is a old and boring subject in computer science, with a history of over 50 years. With the major development and major breakthrough in mobile internet, the internet of things, big data, cloud computing and other key technologies, all hardware on the planet may become intelligent and artificial intelligence revives. Neural networks, on the other hand, are an adaptive nonlinear dynamic system composed of a large number of interconnected basic neurons elements. The system can adapt to the environment on its own, summarize the rules, and finish certain kinds of arithmetic and do some recognition or process controls. The development of artificial intelligence and neural networks are based on many technologies, two of which are big data mining and high speed accessible communication networks.

According to analysis, the data produced every one year since the beginning of the 21st century is far more than the total of the past 2000 years combined. With the development of storage technology and the decrease of the storage cost, the data produced by the human being can now be stored at a low cost and, through the high-speed communication networks, they are also easily accessible to all. Meanwhile, data mining technology is now become maturity. Therefore, based on the massive amount of data, massive knowledge can be formed through data collection, arrangement and analysis. The development of intelligent technology will enable smart machines, with the help of massive data, to compete with or even exceed human wisdom.

With the rapid development of technologies like data storage, data mining, and artificial intelligence, the communication network and the service and experience it provides will also go through significant changes.

Firstly, the massive status data created by network nodes, plus data mining technology, allows the communication network to feel the change in the traffic and the network status. Meanwhile, with the help of the arithmetic of artificial intelligence, the self-organization, self-adaptation and self-improvement of the communication network can be realized. It will be unnecessary for human beings to do the operation and maintenance for the network anymore.

Besides, to better satisfy the needs of Intelligence Ubiquity, the communication network are not only responsible for carrying the information, data and the content, but also for integrating the data and the content. The network node, data and content will become a whole one.

Finally, we can image the more distant future. The network stores massive data of production and behavior. Through data mining and artificial intelligence, the network can understand itself better than humans and can predict individual behavior with the help of big data. At last, the network can be the assistant of all human beings to offer help in everyday life and work.

5 Vision of Telecommunication 4.0

The current communication system cannot meet human being's need for Information Consumption, Sense Extension, and Self-liberalization. Hence, a communication system featured with Information Ubiquity, Sense Ubiquity, and Intelligence Ubiquity is needed, as shown in Fig. 6.

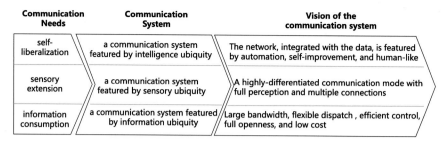

Communication Needs	Communication System	Vision of the communication system
self-liberalization	a communication system featured by intelligence ubiquity	The network, integrated with the data, is featured by automation, self-improvement, and human-like
sensory extension	a communication system featured by sensory ubiquity	A highly-differentiated communication mode with full perception and multiple connections
information consumption	a communication system featured by information ubiquity	Large bandwidth, flexible dispatch , efficient control, full openness, and low cost

Fig. 6 Vision of the communication system

To support Information Ubiquity, the communication network should be featured with large bandwidth and low cost, support efficient and flexible deployment, be capable of delicacy management for traffic and to realize openness.

To support Sense Ubiquity, the communication network should support all kinds of perceptions and massive connections, as well as a dynamic communication mode that is highly differentiated.

To support Intelligence Ubiquity, network should be highly intelligent, and it can realize automation, self-organization, self-adaptation, and self-optimization, support the deep integration between the network, the content and the data.

Generally speaking, in order to satisfy the 3rd–5th level needs in the model of communication needs, the core vision of the future communication network should be an automatic and self-optimization network which can support large bandwidth, low cost, full perception, multiple connections, flexible dispatch, full openness.

Chapter 3
The Main Contents and the Architecture of the Telecommunication 4.0

The core of Telecommunication 4.0 is the combination of the IT industry and the CT industry. Telecommunication 4.0 consists of two main elements: Software-defined networking (SDN) and decoupling of hardware and software.

Software-defined networking (SDN) is an approach to manage network services through software. As in the IP field, unlike traditional approaches, SDN uses a centralized control and opened network architecture to efficiently schedule the entire network resources, providing the openness of network connectivity.

The decoupling of hardware and software refers to unlink software and hardware. The hardware is used as the underlying general platform and the software is used to realize network functiones. NFV, namely network function virtualization, is the core technology to realize the decoupling. NFV is used to decouple traditional network equipment and hardware to create networking services via the common server and the technology of IT virtualization. Ideally, the utilization of NFV will improve the efficiency of management and maintenance, enhance system flexibility.

The components of the network are functional nodes and networking connections. The functional nodes enable the network services, while the networking connections link those functional nodes. NFV and SDN are the core technologies of Telecommunication 4.0, with the NFV used to apply software and the SDN used to accommodate the software. The synergy of these two technologies is Telecommunication 4.0.

1 The New Era—The Advent of the SDN

The cornerstone of the traditional IP network is the routing protocol. For example, the Open Shortest Path First (OSPF) computes the shortest path tree for each route using a method based on Dijkstra's algorithm, instead of computing the shortest path tree for topology and traffic combined, as shown in Fig. 1. Thus, a flood of traffic to some links often occurs but the whole network is maintained or even under

© Springer Nature Singapore Pte Ltd. 2018
Z. Li, *Telecommunication 4.0*, DOI 10.1007/978-981-10-6301-5_3

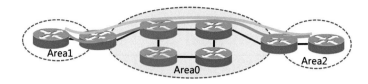

Fig. 1 Optimal path plan is not the optimal plan for traffic

light load. It often deteriorates the quality of the links of the network. The alternatives, such as the MPLS, are not deployed on a large scale due to the complication of the control and maintenance.

To ensure the quality of the network, most of the service providers principally keep their networks under a relatively light load. For instance, they will expand the capacity when traffic peaks reaches 70%. According to statistics, the average utilization rate of a network is approximately 30–40%. The scale of network service providers will be promisingly growing exponentially, considering the global internet traffic will increase ten times from 2014 to 2019. If network servers cannot utilize the network efficiently at that time, the cost will become unaffordable.

Fortunately, Google announced that its IP backbone network connecting the global data center had applied SDN in April 2012, which provides the company a bonanza in savings and efficiency. Such an attempt had enabled the utilization rate of the expensive undersea cable network bandwidth to rise from 40% to nearly 100%, as shown in Figs. 2 and 3. This first successful application for the SDN technology on a large network has shown its bright future for the world of network engineers.

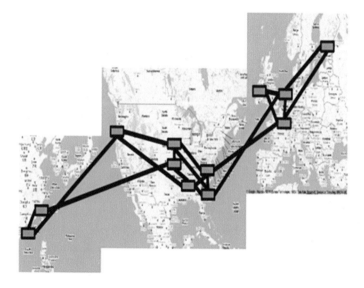

Fig. 2 Google IP backbone network deploys SDN

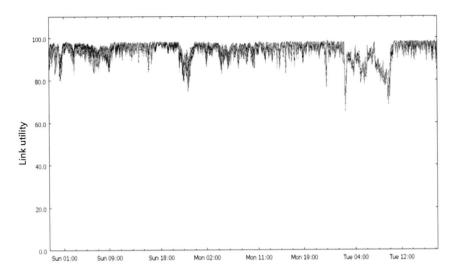

Fig. 3 Utilization rate of bandwidth is near 100%

The successful practice of the SDN not only brings potential growing revenues to the industry, but also supports the future innovative attempts in developing new business services.

The subscription of traditional networks requires customers a relatively long time. For example, the subscription of the national MPLS VPN service will go through a series of processes: application filing, application confirmation, internal order, network departments' exploration in provinces, IP network configuration, network coordination and etc., which usually takes one month.

SDN, however, provides network services through APIs. The services provided cover a wide range and thus customers can choose any virtual network service based on their needs. If a customer intends to subscribe to the MPLS VPN service, the customer can apply for it with such information: the network node, the link bandwidth between nodes and the requirements of the service quality. The customer then can immediately enjoy the service as long as the applied resources are available. This simple process saves time for internal coordination and manual configuration, and it enables customers to experience speedy online subscription.

Considering all its merits, the SDN is regarded as the future network architecture and the emerging area in the industry.

1.1   SDN, Starts with OpenFlow

The SDN originated in Internet companies, and its architecture has been improved throughout its development.

In 2006, the research team led by Prof. Nick McKeown in Stanford University, Clean Slate, released the OpenFlow Protocol. Compared with traditional network devices, OpenFlow separates the control and forwarding into centrally controlled servers and switches based on forwarding table. OpenFlow, acts as a protocol, enables the communication between the servers and the switches.

OpenFlow protocol is just a beginning. The separation of the control plane from the forwarding devices allows for more sophisticated traffic management, then what about the upper layer services? And whether this can be controlled through software? For these questions, Nick puts forward the SDN officially.

In 2009, the version 1.0 of the OpenFlow protocol was released, which is the milestone for OpenFlow due to this version's commercial property. In 2011, the Open Networking Foundation (ONF) was officially funded by prominent companies and organizations, such as Stanford University, Facebook and Google. The new development of the standard was managed by the ONF, and this foundation approved OpenFlow version 1.1, 1.2 and 1.3 successively. The three-layer architecture of the SDN is widely acknowledged.

In July 2012, VMware acquired Nicira, a company focused on software-defined networking (SDN) and network virtualization, for $1.26 billion. This acquisition offers more clues about the bright future of the SDN, although the acquisition costs is not the highest in the internet and telecommunications industry. Nicira was founded by Martin Casado and his fellows, who were in the Stanford research team. At the time of the acquisition, Nicira was founded only 6 years with less than 100 staff. This acquisition triggers the transition from the traditional IP network to SDN.

With constant researches and explorations, the architecture and the main idea of SDN have been widely recognized. The three-layer architecture of the SDN is comprised of the application layer, the control layer and the forwarding layer. And SDN is an approach to separate controllers and forwarding devices, centralized control and provide network services with standard interfaces. However, SDN still can be defined broadly and narrowly.

The narrow definition of the SDN is based on the original SDN concept and the OpenFlow protocol proposed by Stanford University. It uses OpenFlow as the southbound protocol between the controllers and switches, and it also requires forwarding devices to support OpenFlow flow table forwarding. The application layer defines the network with standard northbound interfaces to control and coordinate the network flexibly.

In light of the deployment and the advances of the existing networks, in its broad scope, OpenFlow is often considered as one alternative of the southbound protocols of the SDN. Other protocols can also realize centralized control or management of networks by upgrading software or even using existing routing strategies. Broadly, the SDN architecture is three layers: the application layer, the control or management layer and the forwarding layer. This architecture is also capable of providing northbound interfaces in the application layer to the operation system.

As Shakespeare once said, "There are a thousand Hamlets in a thousand people's eyes": Due to different practical experience, SDN developers differ on the opinions of the SDN. Nonetheless, the basic concept of the SDN has been widely recognized, and it inspired future network builders.

1.2 SDN Architecture

As shown in Fig. 4, SDN consists of the application layer, the control layer and the forwarding layer.

1. The Application Layer

The application layer is the key to the commercialization of the SDN. With an open network architecture, a variety of applications can analyze and coordinate basic networks to provide speedy, flexible and quality services and enhance the efficiency of the utilization of resources and the ability of virtualization.

The current network enables the application layer to function without knowing the complex structure and concrete realization of the network. The application layer needs to define virtual network models, which are adaptable to different network scenarios. For instance, a model of the data center concentrates on describing virtual networks of operating systems or individual renters, while a model of the WAN focuses on network connections, the utilization rate of the bandwidth and the network quality. The application layer connects the control layer with northbound interfaces, such as HTTP protocol. The application layer models will be a focus in the future standardization of the SDN and are being defined in organizations such as ONF and IETF.

2. The Control Layer

Cluster servers, namely SDN controllers, make up the control layer. The main functions of the controller are topology management, path computing, traffic computing and management of flow tables, southbound and northbound interfaces, blogs, alerts, security and reliability. Controllers complete the process of the

Fig. 4 SDN architecture

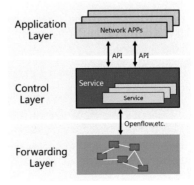

transformation of virtual models to actual forwarding. Controllers first receive resources and deployment applications from the northbound interfaces, then transfer applications into achievable forwarding strategies and finally deliver them to the forwarding layer through southbound interfaces. The control layer seeks to avoid the complication and diversity of physical networks and offer network information, such as operating states and network resources, to the application layer. At present, southbound interfaces can be divided into two categories: network states and configuration protocols, such as OVSDB, Netconf, OF-Config, and XMPP; router/information forwarding interfaces, such as OpenFlow, PCEP, IGP, BGP, XMPP, etc.

Open source controllers are the major driving force of the SDN industry. Currently, the mainstream open source controllers are the OpenDayLight Project launched by IBM and Cisco. The alternatives include ONOS, Open Contrail and Floodlight.

3. The Forwarding layer

The forwarding layer provides flexible and programmable capacity to forward traffic. This layer can be physical hardware or virtual software (i.e., virtual switches) based on actual needs. Physical hardware usually has better forwarding performance while virtual software is more flexible in deployment.

The forwarding layer forwards the data according to the forwarding table delivered by the control layer. It also encapsulates data via VxLAN, MPLS, GRE and etc.

The OpenFlow protocol requires forwarding devices to forward data only after they have matched multiple conditions. This mechanism changes the fixed line of traditional forwarding devices, offering more flexibility and possibility for forwarding. It is also the reason why Stanford University creates OpenFlow protocol. Since the used x86 server is equipped with ample entries for storage, such a mechanism is achievable easily in virtual switches. However, its deployment on hardware switches has not been commercialized due to lack of support (currently only TCAM) and a small scale of entries. Although OpenFlow1.3 attempts to deploy a multiple-layer-forwarding table, its commercialization still failed due to configuration limits for every single entry under the multiple-layer-forwarding table.

New chip manufacturers now seek to design chips suited to the OpenFlow protocol forwarding mechanism. Such an attempt is revolutionary in the hardware chip industry.

4. SDN Utilized in Data Center

Figure 5 shows the virtual diagram of a Data Center.

This model shows that a tenant's network includes three basic components: routers, subnets and interfaces. It also consists of value-added services, such as load balancer, firewall, IPSEC/SSL and VPN. Besides, access strategies, access IP and bandwidth demands are reflected in this model.

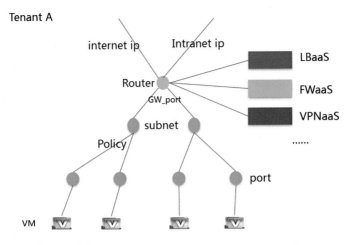

Fig. 5 Virtual diagram of data center

When a user defines a model, the application sends this model's information through Restful with northbound interfaces to SDN controllers. The controllers will allocate IP addresses for virtual machines and hosts, divide subnets, set firewalls, balance network load and generate forwarding table. At last, the controllers forward the table to the forwarding layer via the OpenFlow protocol. Subnets will be sort out with VxLAN ID, and security strategies will be delivered to firewalls through forwarding devices.

1.3 The Application Scenarios of SDN

The SDN architecture brings huge potentials to open networks and improve resource utilization. It can be deployed in the Cloud Data Center, WAN, Transport Network and Mobile Core Network.

1. VPC Service Provided by Cloud Data Center

For the traditional cloud data center, storage and computing resources can be easily virtualized and render services based on needs. Nonetheless, network virtualization can hardly be commercialized due to various reasons. For example, its devices are limited by VLAN, VRF and virtual firewalls, for they have no virtual network models to pool resources and deploy resources flexibly.

VPC requires the networks of the cloud data center to provide self-service for tenants to subscribe, configure and manage isolated networks, as shown in Fig. 6. Specifically, the services include four categories: the immediate subscription of virtual network services; the self-configuration of virtual network services, such as defining subnets services, network segments, access strategies between segments;

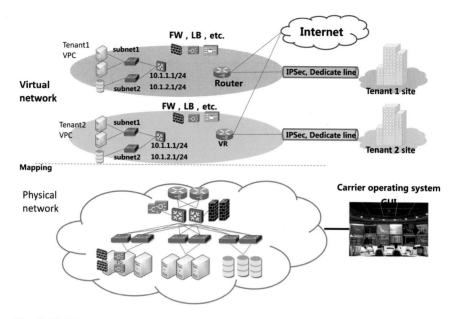

Fig. 6 VPC for multiple tenants

self-management of virtual network services, such as monitoring traffic and topology, isolated networks for each tenant.

SDN network uses VxLAN to expand the scale of the two-layer network, and it realizes the horizontal expansion of network elements with virtual routers and virtual firewalls. With the Neutron model from the OpenStack project, users can define various services at their discretion, including routers, networks, subnets, outside networks, firewalls and load balancers.

The cloud Data Center is the major commercial scenario for SDN, and VPC is the most promising area of SDN.

2. Intelligent Traffic Scheduling of WAN

With the implementation of the 'Broadband China' strategy, HD videoes, 4K videoes and other Internet-based businesses have outperformed the network expansion. Consequently, link congestions occurred in the backbone network, provincial network, metropolitan area network and data center, resulting in packet loss, time delay and unsatisfactory experiences for the clients.

By SDN, the WAN senses the topology, routing and traffic information of network resources to control the network flow, as shown in Fig. 7.

When the software perceives link congestion, a traffic-scheduling algorithm is adopted to assign network flow to more reasonable paths, which guarantees users' service experience and improves the network utilization. For VIP clients, high-quality services can also be provided, such as end-to-end bandwidth

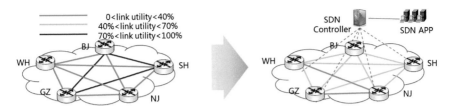

Fig. 7 WAN traffic balancing with SDN architecture

assurance, flow synchronization among data centers, CDN back-to-source flow and mobile backhaul network flow.

The development of the SDN in WAN is mainly driven by intelligent traffic scheduling. The SDN architecture can be utilized in the Cloud Data Center, WAN, Transport Network and Mobile Core Network.

3. SDN Meets the Needs of Speedy Subscription

Usually, it will take months for group clients to coordinate resources to subscribe network services in different provinces. This time-wasting process cannot meet the demands of the industry.

However, SDN enhances network capability and deploys resources flexibly. Clients can define virtual networks depending on their own demands, as shown in Fig. 8. For instance, clients can choose locations of network links, bandwidth between links, specific demands on service (such as time delay and packet loss) and different quality standards (four categories: average, copper, sliver and gold). The service platform will notice controllers immediately after receiving applications to ensure the subscription of the service for the user and the isolation from other users. This automatic process just takes a few minutes.

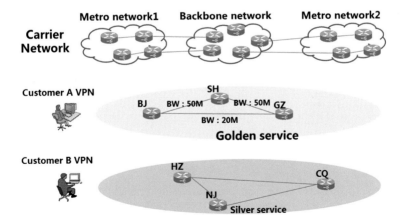

Fig. 8 SDN in VPN network of group clients

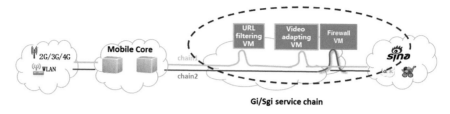

Gi/Sgi service chain

Fig. 9 SDN in Gi/Sgi

4. SDN Generates a Flexible Service Chain

Currently, there are many kinds of Gi/SGi interface traffic strategies. Among them, the HTTP service relies on content cache and page adapter to access the internet, while the video service relies on the video optimizer and protocol optimizer.

For the traditional network, service providers find it difficult to pinpoint every single path of the data. Thus, all data streams have to go through the value-added service platforms, including the content cache, page adapter and video optimizer. This process brings difficulties to the network industry due to its high cost and limited capability.

Business chain is one of the problems that SDN can address. There are two reasons: the forwarding layer of SDN perfectly matches to the business chain: the centralized controllers simplify the distinction of different services. By implementing differentiating strategies and path arrangement of HTTP and video on the application layer, different network traffic goes through the exact functional elements based on needs, as shown in Fig. 9.

1.4 Is SDN Mature Now?

With years of experiment and exploration, SDN has developed rapidly and been commercial deployed. The most salient feature of the SDN is the uneven development in different areas such as service provider cases and plans, chip and open source. In many scenarios, SDN has been commercialized, while others have not.

1. Service Provider Cases and Plans

Since 2012, SDN has been an emerging area, and its maturity curve is shown in Fig. 10. And the deployment of SDN in the Data Center is mature now. By the end of 2015, SDN had been deployed in national and international service providers, large-scale Internet Contents Providers, banks, oil companies, governments and universities.

Clouded sites refer to the next generation SDN&NFV data centers, for example, core network computer rooms where SGSN or GGSN are placed and fixed network computer rooms where BRAS or SR are placed. Clouded sites bear the capacity

Fig. 10 SDN maturity curve

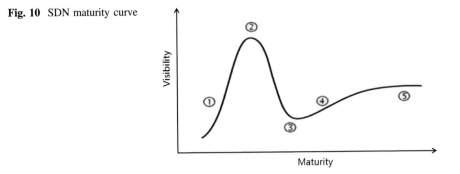

with general servers, virtual networks and SDN networks together. A few advanced server providers abroad have tried to commercialize the vBRAS sites, yet the commercialization has not been completed.

At present, major data communication equipment manufacturers and innovative enterprises prioritize the areas of data centers and clouded sites. Manufacturers have many plans, and most plans are mainly based on OVS+ODL+OpenStack to provide end-to-end solutions within data centers. One interesting phenomenon is that the model of the utilization of SDN systems is becoming more mature and generating more benefits for the community. It is well known that the standards of traditional network products are relatively high. That is to say, a mature product may be an extremely complicated system, including hardware, software and even chips. Also, it will take a quite long time to examine bugs and maintain the product. Thus, there are few manufacturers currently in this industry, and other new companies find it difficult to fit in. However, under the open source model, small-and-middle-sized companies can rely on the advances of open source to catch up with the giants. This new model may have profound influences on the profit model of this industry.

SDN is also widely used in the IP WAN, including the large-scale Internet content providers (i.e. Google and Tencent) and the data center operators who use it to enhance efficiency. However, from the perspective of the entire industry chain, SDN is not mature yet.

The transport network mainly focuses on mobile backhaul network. The three major telecom operators seek to upgrade their existing backhaul networks, and the industry chain is becoming more mature. In WAN and the transport network, manufacturers tend to forward strategies by reforming network management. Andin forwarding protocols, they tend to upgrade the existing protocols to enhance the network capacity with the broad SDN architecture.

2. Chip, New Opportunity to Boost the Industry

SDN chip providers are the critical factor in determining the SDN development based on OpenFlow. New chip manufacturers, such as Centec, MTK, Cavium and Mellanox, have actively promoted the OpenFlow chips to meet the market demands and catch up with traditional companies. However, traditional manufacturers tend to develop the OpenFlow adaptive technologies on the basis of existing chips.

3. **Open Source to Fuel SDN Development**

The SDN-related open-source community has actively promoted the development of SDN. In terms of the layer of deployment, it mainly includes operating system OpenStack, SDN controllers such as ODL, ONO, Open Contrail, Floodlight and RYU and the forwarding layer OVS. ODL and ONOS are the open source of controllers, OpenStack is the open source of northbound interfaces and the OVS is the open source of forwarding modules.

OpenDaylight is an open SDN controller project, which is authorized by EPL and led by Cisco, Brocade, HP and other IT companies. Up till now, it has gained momentum in the community. Over 800 representatives participated the ODL summit this year, and many major IT companies are summit members, including AT&T and Tencent.

Opendaylight seeks to build a development architecture of the open SDN controller and its responding open ecosystem. Opendaylight has already released three versions of Lithium, containing over 60 sub-projects and over 2,400,000 lines of code. Opendaylight provides abundant northbound and southbound interfaces of the SDN controller, such as RESTCONF, Netconf, OVSDB and OpenFlow, and it meets the demands of the data center. Besides, the architecture based on Yang model can well decouple functional modules to suits various forwarding devices.

Opendaylight has become the standard controller of SDN in data centers, since many companies such as Huawei, Brocade, Ericsson and ConteXtream are developing commercial SDN controllers based on Opendaylight.

ONOS is an open SDN controller projected authorized by Apache[1] and is led by an NGO On.lab. ONOS focuses on service provider scenario, and its members include AT&T, NTT and China Unicom. Huawei is the major promoter and contributor of ONOS, and it tries to compete with other IT giants such as Cisco and Brocade by leading ONOS.

ONOS seeks to build open and specific carrier-class SDN controller platforms. Its full-distributed architecture provides a reliable platform to avoid single point's failures. The ONOS architecture also offers a high throughput and low latency. It claims that it is able to process 1000,000 concurrent requests while limiting time delay to 50–100 ms. ONOS has released the 3rd stable Cardinal version, containing a dozen sub-projects and over 350,000 lines of code. It provides multiple solutions for carrier access networks and WAN traffic scheduling scenarios. Recently, AT&T has deployed ONOS in its CORD project. ONOS will become a desirable choice of SDN controllers in the company's future development.

Both ONOS and Opendaylight are members of the Linux Foundation. These two open SDN controller projects will cooperate much in the future.

---

[1]Apache is the world's most used web server software.

SDN Standardization is one of the priorities of its development. It mainly involves ODL, ONOS, OpenStack and OVS.

ONF puts forward the concept and architecture of the SDN, and it is the drafter and maintainer of OpenFlow. It has set standards for SDN architecture, northbound interfaces and network models. ONF contains four areas: carriers, businesses, standards and markets. There are a dozen of working groups, covering architecture, forwarding protocol, carrier-class SDN, northbound interfaces and tests. ONF currently has 134 members, including carriers, equipment manufacturers and chip companies.

IETF is the main organization of the Internet, and the birthplace of Yang model. It is also the drafter of the SDN basic protocol. The specific network model Yang has been preliminarily completed by IETF, who is working with other open source organizations to extend other models now.

The application of SDN has outperformed the development of carriers. Thus, SDN may bring opportunities to the future of the Internet.

1.5 Technical Challenges Faced by SDN

1. Centralized Control to Cope with Large-scale and Complicated Network Challenges

It is difficult for centralized control methods to cope with large-scale networks. Giant cloud computing resource pools need handle thousands of physical switches and hundreds of thousands of virtual switches. A large-scale telecom network also has hundreds of thousands of types of network devices and nearly 500,000 routers. These large-scale centralized control, configuration and fault locations pose challenges to the application software.

Device faults and upgrading of the network will render tremendous changes to network environments due to their complexity. It tests the algorithm of SDN software to see if it is rigorous enough to handle the situations rapidly and deploy resources effectively. This is another challenge for SDN application.

There are blanks in the SDN solutions to large-scale tests, requiring the entire industry to fill.

2. Security and Reliability of Centralized Control

The centralized control method of SDN poses potential threats to the security of the controllers. It requires isolated protection and backup mechanisms to ensure security. This mechanism includes the security issues of the controllers, the connection between controllers and the application layer and links between controllers and forwarding devices.

For the SDN network, dysfunctional controllers may be catastrophic for the forwarding of the network. The way to avoid the dysfunction of controllers by separating networks physically and the way to track back to traditionally distributed networks once dysfunction occurs are the top agendas for the industry.

3. Smooth Transition to SDN

It is also a problem for SDN technologies to satisfy the needs of diverse devices of different manufacturers and realize the coordination of all the businesses.

4. Interface Standardization and Interoperability

Northbound and southbound interfaces' standardization is the only way to connect the layers of SDN. But the southbound interfaces are not standardized yet. For example, although most of the manufacturers support OpenFlow1.3 in data centers, they cannot connect with each other due to the differences in forwarding tables of OpenFlow. As for northbound interfaces, SDN users tend to design their own models, and the designed models are not identical.

The development of eastbound and westbound interfaces is still at a preliminary phase and no standards have been released yet. One goal of SDN is to realize end-to-end virtual network services, that is, to pool diverse kinds of network resources such as the transport network, the IP backbone network and data centers. In the foreseeable future, specific controllers will be implemented in cloud data centers, IP backbone networks, MAN, the transport network and mobile core networks. The coordinate of these controllers necessarily involves the standardization and interoperability of westbound and eastbound interfaces.

Overall, the challenges faced by SDN include the technical ones, and also ones from operating models, operating architecture, companies' architecture and even the internal system. These factors beyond technologies will bring more profound influences on the development of SDN.

2 NFV Opens the New Era of Software and Hardware Decomposition

Nowadays, the network of operators consists of a large number of communications devices, with routers and switches included. One of the important features of these traditional communications devices is that their functional software and hardware devices are tightly bound. Because of different function requirements, these hardware also appears to be in various shapes. See Fig. 11 for example:

BRAS   MME   HSS
(Broadband Remote Access Server) (Mobility Management Entity) (Home Subscriber Server)

Fig. 11 Various traditional communications devices

These different devices are deployed into different machine rooms of the communications network. When the operator is deploying a new service, new devices will be introduced into the machine room, and the old network devices will be gradually replaced. Thus, new devices will always be added when operators intends to adopt new services. The frequent replacement of hardware devices usually brings about high cost input. The cost is not only generated by device replacement, but also by the labor cost of operation and maintenance and the storage space and power supply of these new devices. According to the GSM 2015 report, there were 3.6 billion global mobile subscribers in 2014, and that will probably increase to 4.6 billion by 2020. With the increased number of accessed subscribers, the cost for the operators is surging. The report shows that the investment of global operators was 1.15 trillion dollars in 2014, and it is estimated to increase to 1.4 trillion dollars in 2020. With the subscribers increasing, the asset input and energy costs of the operators shoot up which means that construction investments will be enlarged. Such kinds of service construction modes will bring about huge cost pressures and are becoming increasingly infeasible.

In addition, proprietary hardware devices often requires long terms of development cycle, which results in delay of replacement of the communication network. LTE, as an example, took nearly 10 years from putting forward a concept to determining standards. Later on, the hardware device of LTE was gradually designed and produced. Generally, for a proprietary communications device, 12–36 months are needed for hardware integration design of the CPU, memory and network card, and then about 6–12 months for the development of assorted software. Considering the circle of early requirement surveys and hardware production, it takes 2–3 years to develop new proprietary hardware devices. The long term of the development cycle makes traditional communications device replacement rather slow, which leads to the operator's slow response to new services. Nevertheless, because of the demand of technology and service innovation, especially the rapid development of the internet, subscribers have huge and frequent demands for new services. Therefore, it has become a critical contradiction in the traditional communications network between subscriber's urgent demand for new services and the operator's slow response to these new functions.

By contrast, flexible and open development modes of the internet provide fast response to new services. According to statistics, it often takes 100–200 days at most for a new internet service to go online see Fig. 12. The official report of Apple Inc. shows that in 2012 alone, Apple Store[2] uploaded as many as 869 applications every day. From the end of 2011 to the beginning of 2012, with the success of YMS, the first domestic tax-hailing app of the mobile internet, close to over 30 similar tax-hailing apps went online in just several months. By contrast, due to the restraint of device's upgrade costs and speed, it seems that the operator provides very slow service update. This result in severe impact on the revenue of the new

---

[2]Apple Store is a retail chain store operated by Apple Inc., mainly selling computer and consumptive electronic products.—Editor's note.

Fig. 12 Examples of internet service going online

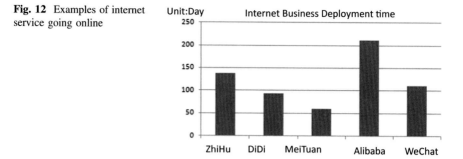

communications network service, and is confined to a technological innovation of the new service landscape.

Under the dual burden of cost and service innovation, operators should find a new way to change the current mode of the communication network. The concept of NFV is raised up and widely accepted by operators under such a demanding background.

NFV, at first, was launched jointly by 13 mainstream operators (American Telephone and Telegraph Company, China Mobile, British Telecom, Telefonica of Spain, and etc.) in 2012 at ETSI (European Telecommunications Standards Institute), and NFV white paper was issued in the same year. Since the launch of NFV, the concept has been widely emphasized. It has gradually become an industry consensus to drive future rebuilding and development of the internet based on NFV technology.

From the white paper in 2012, to the recognized first year of commercialization of NFV in industry in 2015, NFV has realized an important leap from concept to practice. Actively driven by operators, the industry chain gradually comes into being, standards and OSS are gradually improved, and the growth of the industry is inspiring.

2.1   By Introducing IT, NFV Subverts the Traditional Implementation Mode of CT Devices

In recent years, the most attractive technology should be the virtualization technology. Actually, such concept of virtualization was first proposed as far back as the 1960s. In the last 10 years, virtualization technology based on Intel and ARM hardware has developed rapidly, laying a concrete technological foundation for virtual cloud computing. In August 9, 2006, Eric Schmidt, the CEO of Google, first put forward the concept of "cloud computing" in Search Engine Strategies. In the same year, Amazon launched its cloud computing service, AWS (Amazon Web Services), providing enterprises with an IT infrastructure service in the form of the Web service, namely clouds service. Since then, cloud computing services based on IT virtualization technology have aroused wide attention and permeated into all parts of people's life. Today, Alicloud, Baidu Cloud and other various cloud service suppliers emerge one after another, and OpenStack, a cloud management project, as well as start-ups based on OpenStack are growing vigorously. Cloud computing and cloud services are the general trend of IT development.

Virtualization cloud computing technology realized the pooling of resources. Large amounts of general-purpose hardware are pooled into virtual resources and are flexibly invoked by different subscribers. Cloud computing technology fundamentally decreases the construction and maintenance costs of the IT industry, increasing the flexibility, agility and feasibility of the industry. The technology changes the IT industry from the root, laying a technological foundation for rapid growth and maturity of the IT industry.

The success of virtualization technology in the IT industry also arouses the attention of the CT industry. Because of the increasing demand for upgrade the communications networks and the frequent need for service innovation and replacement, operators face increasingly growing cost pressure. Virtualization cloud computing technology supplies virtual resources with low cost and high flexibility, which is the hope of operators to solve the problem of their communications devices. The concept of Network Function Virtualization hereby comes into being.

As to the means, NFV can hopefully integrate existing network device functions into standardized IT general devices in the form of virtual network function software, by using IT virtualization technology. Compared with traditional proprietary hardware devices, after employing NFV it can effectively avoid the replacement of proprietary hardware devices when introducing new services. In NFV, all network elements are deployed into general-purpose IT devices in virtual software form. Therefore, deploying new services only means deploying new service network functions on general-purpose IT devices in the way of virtual software, while there is no need to change the overall hardware devices at the bottom correspondingly.

According to the NFV white paper, "NFV aims to leverage standard IT virtualisation technology to consolidate many network equipment types onto industry standard high volume servers, switches and storage, which could be located in Datacentres, Network Nodes and in the end user premises."

Compared with the traditional communications network based on proprietary hardware devices, NFV can bring the following benefits.

1. **Decreasing the costs of the hardware**

Because the physical entity and the network function of traditional proprietary hardware devices are coupled into a special device, their costs are relatively high. Taking CSCF (Call State Control Function), HSS (Home Subscriber Server) and SBC (Session Border Controller) which have the capacity of a maximum of 1 million concurrent subscribers as examples, generally, their prices are from a few million up to 10 million RMB, of which nearly 30% are design and manufacturing costs of the hardware. Comparatively, owing to the large-scale application in the traditional IT industry, the general-purpose IT devices decoupled from the virtual network functions can dramatically cut the device costs. Generally, the cost of the general-purpose server devices in the current NFV deployment is 20,000–30,000 RMB.

Meanwhile after introducing NFV, the upgrading of services is the upgrading of virtual software, not that of hardware devices. So the life cycle of hardware devices won't impact the life cycle of network functions.

2. Decreasing the innovative cycle of traditional operators and speeding up the commercialization of services

If a new hardware platform is about to be designed, it takes 1.5–2 years for a traditional communication device to go from designing hardware to developing software. While, for a new application of mobile networks, its development only takes around 3 months, and its software upgrading takes even less.

Thanks to the NFV, the innovation and commercialization of communications services changes from traditional hardware development to software development. The threshold for developers is significantly lower, and the development integration and deployment is obviously quicker. The virtualization of network functions can notably decrease the preparation cycle of communications services, promote services to go online and improve innovative capacity.

3. Enlarging or narrowing service capacity rapidly

In the communication systems, "tidal effects", which means that in the same place, the traffic flow will have fundamental changes at different times is quite common. For example, in CBD, the traffic flow will surge during the business day, while it decreases in the evening. Traditional communications devices are deployed according to the calculations of traffic models. The number of deployed network elements is fixed, and so is the subscriber capacity they carry. Therefore, there is rate of call loss in communications networks, which means that when the number of subscribers surpasses the network capacity, the network will refuse certain calls. The fixed deployment mode of network elements in traditional communications networks is contradictory with the demands of subscribers featuring tidal effects see Fig. 13. It leads to frequent call loss during heavy traffic times of communications devices, influencing subscribers' experience. In the meantime, the devices may be idle during idle traffic times, which leads to huge waste of the costs of devices, energy and even operation and maintenance.

Compared with the fixed deployment model of traditional communications devices, NFV can bring dramatic flexibility to service deployment for operators. Operators adjust the number of network elements of online services in accordance with the current network demands. When facing the peak time of subscribers, operators can increase the number of the virtual network functions. In the evening

Fig. 13 Tidal effects

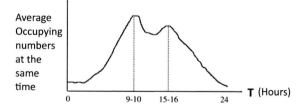

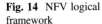
Fig. 14 NFV logical framework

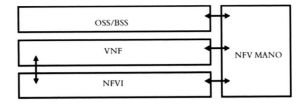

when the number of subscriber is plummeting, operators can reduce the operating virtual network elements and integrate the operating network elements into several hardware devices so as to decrease the energy consumption of the whole networks.

2.2 Starting from the Concept, Reference Framework Is Gradually Developed for NFV

Similar to any other new technology, the concept of NFV has experienced a deeper level of research from its birth to its formation of a feasible reference frame. From the founding of ETSI NFV ISG (European Telecommunication Standards Institute Network Function Virtualization Industry Standards Group), it embarks on the research and discussion of the demands and framework of NFV. The NFV logical framework can be simply divided into the following several parts see Fig. 14.

Figure 14 simply shows the 4 parts of the NFV logical framework, including MANO (Management and Orchestration) for overall orchestration, control and management, NFVI (NFV Infrastructure) which network elements depend on, VNF (Virtual Network Function) and OSS/BSS (Operation Support System/Business Support System). Such a logical framework simplifies the internal specific details of all parts, explicating the NFV's implementation mode from a logical perspective. With the deeper research on NFV and richer NFV deployment experiences, such framework has already become the widely recognized actual deployment framework in the NFV field, in which physical entities that all logical parts correspond to are gradually clear. The following part will illustrate all modules and concrete realizations of the framework.

1. **NFV infrastructure (NFVI)**

NFVI includes NFV hardware facilities and virtual facilities. NFV hardware refers to the general-purpose hardware that NFV software operation relies on, including servers, memory devices, network devices, and so on; NFV virtual infrastructure refers to the host operating system and Hypervisor[3] running in these general

---

[3]Hypervisor is also called virtual monitor. It is an intermediary software operating between the physical server and operating system, and can allow several OS and applications to share one basic physical hardware.—Editor's note.

hardware, which provide virtual computing, virtual storage and virtual network service to the upper layers.

NFVI is the foundation of VNF operation. The capacity and reliability of NFVI significantly influenced those of the VNF operating on it. Under the monitoring and management of VIM (Virtual Infrastructure manager), VIM can invoke all NFVI and deploy missions for it; at the same time, all information of NFVI can be uploaded to VIM in time.

2. VNF in NFV

VNF in NFV makes the whole NFV framework achieve real communications network function. In traditional communications networks, proprietary hardware devices are used to form networks. While, in the NFV frame, these proprietary hardware devices are transformed into VNF. In the form of software, these VNFs are operated in the VM (Virtual Machine) or containers which NFVI provides for them. Thus, a similar service function to that of traditional proprietary devices is provided.

Although it appears like that, transforming from proprietary hardware devices to VNF does not simply requires making software in traditional proprietary devices operate in the VM of NFV. The meaning of NFV not only lies on the superficial separation between software and hardware and the introduction of VM technology. It also concerns the division of different parts in CT network elements and its realization by way of Virtual technology in IT. In this process, the capacity of IT devices should be taken into full consideration, and the existing CT network elements should be redesigned in a fine-grained way. It shouldn't be the goal of NFV to rely on a "huge" VM to achieve the whole function of traditional network elements. However, NFV should be the realization of a whole network function by depending on several small and fine-grained VNFC (VNF Component), featuring flexible deployment and management, scalability and high reliability. Such fine grained VNFC contributes to a more flexible deployment of VNFs and more fully use of virtual resources. Small VNFC needs fewer virtual cores and several VNFCs can be deployed into one server so as to make the most of all virtual core resources. Large VNFC, however, needs more virtual cores, and one VNFC often occupies one server. Thus, the rest of the virtual cores cannot be fully used.

3. NFV management and orchestration:/NFV MANO

In the framework of NFV, Orchestrator, VNFM and VIM are collectively called NFV MANO. In the whole NFV framework, NFV MANO plays a key role of management, control and coordination, in which its three components have their own functions respectively.

(1) Orchestrator

Orchestrator is the brain of the whole NFV framework, responsible for accepting the deployment demands distributed by operators and OSS/BSS, and for distributing specific deployment commands to VNFM and VIM. Orchestrator will orchestrate and manage the whole NFV framework and is responsible for the deployment and management of network services. The deployment missions distributed by orchestrator contain the abstract description on network services, and VNFM and VIM will be responsible for making such description deployed on a concrete virtual machine constructed.

(2) **VNFM**

VNFM plays a role of VNF manager in the whole NFV framework. Its main duty is to manage the life cycle of VNF, including VNF instantiation, detection, scalability, termination and so on. In the NFV framework, many VNFMs can be deployed. VNFM and VNF are closely related, so VNF and VNFM are usually tightly binding.

(3) **VIM**

VIM is the manager of deploying NFV infrastructure in the whole NFV reference framework. Its specific jobs contain management, monitoring, allocation of hardware resources and their virtual resources. Although VIM refers to Virtual Infrastructure Manager, the management, monitoring and allocation of physical infrastructure are also the main functions of the VIM, according to deeper research of NFV.

At present, in the NFV industry the widely recognized specific component of VIM is OpenStack. If SDN technology is used to build the virtual networks, SDN controller can also be regarded as a part of VIM. OpenStack is the traditional management component of IT cloud, and in NFV, it can also be used as the manager of the virtual infrastructure. However, to realize the NFV framework, there are special requirements different from the traditional IT cloud, and more requests on OpenStack are also made. SDN controllers are used to allocate and manage networks in the whole NFV deployment. Compared to only using Neutron, the network component of OpenStack, using SDN controllers to control networks is more flexible. According to the actual deployment requirement, more than one VIMs can be co-existing in the NFV framework.

(4) **OSS/BSS**

OSS/BSS is a component to connect NFV with traditional telecommunications networks. It is a supporting system where telecommunication operators can share integrative information resources. It mainly consists of network management, system management, billing, business, accounting and other subscriber's services. After introducing NFV, the new virtual network should be added to the traditional one so as to be controlled. An important connecting component is OSS/BSS. The interface between OSS/BSS and orchestrator will be responsible for delivering Network service orders, and the Orchestrator will transform these orders into the NFV template. Therefore, this interface is of high importance.

2.2.1 Communication Hardware Is Marching Towards Generalization, Developing from CPCI, ATCA to COTS

The hardware of communications networks has experienced a gradually evolutive development process, growing from proprietary hardware of CPCI (Compact Peripheral Component Interconnect) to widely applied ATCA [Advanced Telecom Computing Architecture)] frames. In Communications 4.0, hardware will evolve into integrated, general COTS (Commercial Off-the-shelf), to achieve devices generalization. With the help of different software, different demands can be met so as to adapt to the fast innovation and upgrading needs of telecom networks.

CPCI refers to Compact PCI (Peripheral Component Interconnect), which is a bus interconnecting standard proposed by the PCI Industrial Computer Manufacturers Group (PICMG) in 1994. It is a high-performance industrial bus with the standard of PCI electrical specification. The hardware devices employing the CPCI specification feature with high reliability, hot-pluggable capacity, and better bandwidth. Therefore, it is widely applied in the communications industry, especially in high speed data communications devices, such as routers and switches.

However, with the rapid growth of semi-conductor technology, the shortcomings of CPCI devices are gradually being exposed. First, the chip board of CPCI is narrow. As the chip scale is enlarging, the narrow space can hardly occupy enough chips, which greatly impacts the performance of CPCI devices. Second, the power consumption limitation of the single slot of CPCI devices is too low, with only 45 W (Watt). With the development of chipset technology, the core power consumption of CPU with high performance will be gradually improved, and multi-core technology will be increasingly advanced. Thus CPCI devices' limitation to power consumption will significantly influence its capacity for carrying several cores. Besides, in dual-star switch architecture based on PICMG 2.16, bandwidth of the CPCI single slot is only 1G as a maximum level, which cannot satisfy the growth demand of high speed data communications.

Because of the network's high demands of gateway heat dissipation, single board space, bandwidth, high reliability and system management, PICMG is encouraged to distribute the PICMG 3.0 standard by the end of 2002. Such standard is the core specification of ATCA, identifying all interfaces' definition, electrical property, backboard structure, system management specification, mechanical structure and designing standards, etc.

ATCA refers to Advanced Telecom Computing Architecture, growing out of CPCI. In order to integrate communications data network applications, it provides a cost-effectively, modularized, compatible and scalable hardware architecture. Targeted mainly on telecom operation application, ATCA is shown in a way of modular structure, so as to meet the needs of communications for high speed data transmission and supply telecom operating devices with a highly reliable and feasible solution. Compared with CPCI devices, ATCA boasts a flexible backboard structure, supporting various topological structures, such as star-like, mesh, dual-star and two groups of dual-star structures. Its internal highest data transmission and switch rate reaches 2.4 Tbps,[4] and it can support various switch buses. Meanwhile, ATCA specification reserves more room in its shape design, and the monitoring and management of redundant fans and temperature, to strengthen the heat dissipation capacity of the system. These designs not only effectively guarantee that communications devices can achieve carrier-grade stability which helps devices be effective when facing heat dissipation, but also provide these devices with a strong handling capacity in using multi-core or high speed CPU. As for reliability, the slot, power and fan of the ATCA all adopt dual backup mechanism so that single board failure will not influence the whole case. Moreover, in ATCA architecture, the management plane, control plane and service plane are totally separated. They all adopt a point-to-point structure, effectively improving the platform reliability. Besides, most importantly, ATCA dramatically enhances the compatibility of different manufacturers by using a unified open bus design with the whole network standardization. At the same time, operators can upgrade their services through changing board cards.

With the rapid development of the internet, OTT (an internet application service) has an increasingly obvious impact on core telecom services, and the development mode of the traditional telecommunications industry featuring "high performance, high reliability, and high cost" is gradually questioned by operators, so the concept of NFV comes into being. Operators hope that telecom services can get rid of the vertically integrated mode of traditional proprietary hardware device, and break through the tight binding between hardware and software. It is expected to take the telecom industry out of the "group of smoker" of proprietary devices, and make use of unified and low-cost hardware, to achieve VNF application based on cloud computing virtualization technology.

The NFV architecture based on general-purpose hardware, on the one hand, can effectively decrease the costs of proprietary hardware. On the other hand, it can make sure of the flexibility of the whole telecom architecture. In traditional proprietary hardware devices, hardware and software are closely tied with each other, which brings about high threshold for the industry. Few manufacturers can participate in such industry resulting in high pricing. Operators are bound by several devices manufacturers, so the costs of devices can hardly be cut. In the IT field, however, the costs of those general hardware devices which have been utilized for more than 10 years are gradually decreased, owing to the standardization and mass production of the devices' components. Moreover, the software and hardware are coupled in proprietary devices, which means that operators have to restart customizing, designing, integrating and developing high-cost hardware devices whenever they are upgrading services. The deployment has a long period and is limited by manufacturers. After adopting the VNF, the upgrading of services has become that of software, leading to simple operations, low costs and rapid upgrading. In face of the increasingly fierce market competition, operators can make a great stride in the speed of upgrading services.

---

[4] 1 Tbps refers to terabits per second.—Editor's note.

However, simply introducing the general-purpose devices in the CT industry is not enough. Traditional proprietary devices are designed to have strong packet-processing and forwarding capacities, which cannot be matched by the general-purpose ones. Semi-conductor technology is developing, the CPU capacity is growing, and the multi-core technology is maturing, but general-purpose devices' packet forwarding capacity still cannot totally meet the demands of the network devices of high speed data. There are many solutions targeting the problem. In software, the use of user plane data acceleration such as DPDK (Data Plane Development Kit) can effectively improve the data processing capacity of software. In hardware, the use of SR-IOV to bind certain network card or using PCIe (Peripheral Component Interconnect Express) for pass-through, can make binding gateways provide data forwarding services for certain service, and avoid the loss of forwarding capacity brought by virtualization software. Besides, accelerator cards are widely used to transfer the large amount of data processing works from CPU to the accelerator cards, so as to improve the whole performance. However, the above-mentioned method to solve the performance problems by specific hardware operation will undoubtedly decrease the flexibility brought by the NFV virtualization. For example, the VNF using SR-IOV cannot realize live migration, and the unloaded software using accelerator cards cannot migrate to other hardware devices without the deployment of accelerator cards. They all show that in this period, to apply the NFV based on general-purpose hardware devices, operators have to make a balance between the network service performance and the flexibility of virtualization.

2.3 The NFV "Makes Great Achievements" in Various Fields of Communications Networks

The concept of NFV overturns the whole communications network. From the edge access network to the core network, from fixed network to mobile network, all network functions may be re-upgraded thanks to the concept of NFV. This part will deeply introduce several important scenarios in the NFV deployment practices. These deployment scenarios are firstly selected by operators and manufacturers to serve as an important breakthrough of the NFV practice.

1. The NFV in Fixed Access Network

In traditional fixed access network, the edge devices of users are connected to operators' BRAS through network access devices. Customer premise equipments provide basic network services for them, for example, PPPoE (Point-to-Point Protocol over Ethernet) dialing, firewall and Layer three routing. With the increasing innovative demands of households and enterprises for network services, customer premise equipments need to have more and more functions, and their physical devices are more complex as a result. These physical devices should be

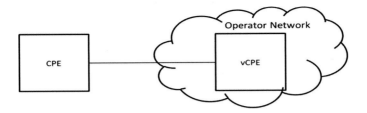

Fig. 15 vCPE deployment architecture

frequently upgraded to add new functions. Meanwhile, because the devices are more complicated, the failure rate and operation and maintenance costs are increasing.

At the same time, a large number of new network services come into being, which brings a huge challenge to the fixed access network. For instance, various household access boxes, such as MI box are snatching the user access, providing richer services for users. These new service modes bring huge pressure to operators, and encourage them to step up the online speed of new fixed access network services, enrich service contents and guarantee service quality.

Under the dual pressure of costs and innovation, operators are encouraged to promote the NFV deployment practice in fixed access network. vCPE (Virtual Customer Premises Equipment) has been regarded by most operators and manufacturers as the most typical case in the NFV deployment practice in fixed access network see Fig. 15. vCPE will deploy relatively complex network functions and new services into the network side instead of user side by way of NFV virtualization, and provide an open platform for the operator's future new services and even third party services. The deployment of vCPE makes traditional fixed access networks more flexible, and users can customize their own networks services according to their demands.

At the same time, virtual OLT (Optical Line Terminal) technology, virtual BRAS technology and other fixed access network virtualization solutions are gradually raised up. Based on the current industry development status, NFV in fixed access networks has become a consensus in the industry, but current technological solutions are not yet mature, and various VNF functions still need to be improved. Especially after introducing NFV, the function distribution of various VNF and the new networking topology should be taken into further consideration.

2. **The NFV in Mobile Access Networks**

With the development of wireless communications technology, mobile access networks are growing mature. From GSM access in the 2G era to 4G LTE access, mobile access networks have experienced a dramatic change in the short period of just 20 years. The mobile access network devices are widely deployed, the access technology continues to innovate, and the number of the supporting terminals is increasing. By January of 2015, the number of 4G base stations of China Mobile had reached 700,000 and that of 4G users reached 0.1 billion.

However, the rapid growth of mobile access networks also brings about the severe problem of costs. Mobile access devices are widely distributed with huge numbers and high energy costs. Meanwhile, the management of distributed devices is growing complicated, which is a huge challenge for the further development of mobile access networks. Facing these problems, the virtualization of mobile access networks is the important solution to get rid of this dilemma.

Take the virtualization of a small cell gateway as an example. With the broad construction of LTE networks, the small cell is taken as a key way to solve the problem of indoor coverage of 4G wireless broadband. Nowadays, small cells rely on the small cell gateway to access mobile core networks. The small cell gateway devices provided by traditional network device manufacturers are high in cost, with over 1 million RMB per device. With the wide deployment of small cells, operators have more demands on the small cell gateway, and decreasing the costs of small cell gateways has become the key demand of the operators. Comparatively, adopting NFV solutions to achieve small cell gateway and introducing virtualization technology to deploy small cell gateway services in the platform of general server hardware will cost around 200 thousand RBM. It significantly decreases the costs of purchasing, maintenance and management of traditional proprietary network devices, and becomes the ideal choice to achieve the small cell gateway.

Meanwhile, research on C-RAN (C-Radio Access Network), a new base station construction structure, is also ongoing in recent years. Through moving the baseband processor away from the station and deploying it into a deeper part of the core, C-RAN can effectively decrease the cost of devices, improve coordination and increase network capacity. In the future, C-RAN is expected to become the mainstream of communications network infrastructure construction see Fig. 16.

Meanwhile, in the aspect of wireless access networks, the AC (Access Controller) virtualization and pooling technology based on virtual technology gradually attracts broad attention. The achievement of the AC virtual pool can effectively cut AC devices' costs, increase their reliability and enhance AC devices' utilization. At the same time, operators can simplify the maintenance and management of AC devices.

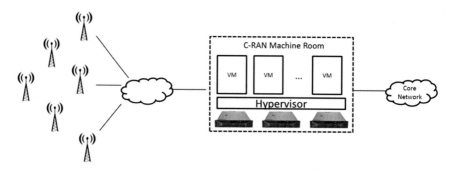

Fig. 16 The deployment architecture of C-RAN

The distributed feature of the mobile access network brings the challenge of high costs and high management complexity. It makes NFV an important solution to its future development. Currently, different virtual mobile access network devices based on the NFV solution have already sprung up. Virtual small cell gateway will hopefully be commercialized in 2016 in China. Field tests related to C-RAN architecture are conducted in many provinces. In the several coming years, the NFV products for mobile access networks will be further matured and will compose virtualized communications access networks together with fixed access networks.

3. **The NFV of Mobile Core Networks**

Mobile core networks, the core of mobile communications, undertake several services, including call connection, billing and mobile management. With the growing development of mobile internet technologies, mobile core networks have been increasingly innovated in the time of 2G, 3G and 4G. There are more and more network elements of mobile core networks, undertaking a growing traffic flow. Core networks have higher demands of decreasing costs and strengthening services scheduling flexibility.

In the NFV industry, the virtualization of core networks is always the focus of operators and manufacturers. Compared with the current vertical application deployment method of the core network, the decoupling between bottom hardware infrastructure and upper layer application based on the NFV can make the bottom network resources be shared by various applications. At the same time, mature network virtualization technologies can help operators flexibly allocate network resources, and achieve automatic deployment, live migration, and automatic scale-up and scale-down of the application according to traffic load, location and time see Fig. 17.

Nowadays, there are many POC (Proof of Concept), field tryouts and deployment cases related to core network virtualization of global operators. For example, in 2015, DoCoMo, a Japanese operator, tried to deploy self-integrated vEPC

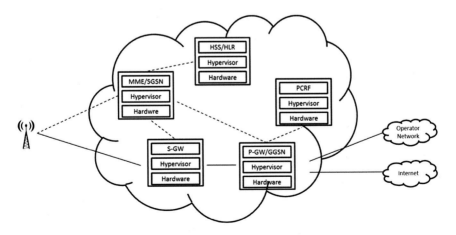

Fig. 17 The realization of NFV in core networks

(virtualized Evolved Packet Core Network) in the network of the Internet of Things. Meanwhile, China Mobile also starts the NFV field trail of vIMS (virtual IP Multimedia Subsystem) as a start for the NFV of Mobile Core Network.

From the cases of many core network virtualizations, we can see that the entire NFV industry has a huge investment on core network virtualization. However, the virtualization of the core network still faces the challenges of upgrading the existing complex communications network, changing the devices resulted from virtualized deployment, deploying software and even achieving the follow-up operation, maintenance and management. The realization of core network virtualization still needs a period of technological accumulation and operation and maintenance upgrading.

4. The Applications of the NFV in Data Centers

Besides providing host co-locating service, traditional data centers also provides the value-added services, such as firewall, load balance and VPN. These value-added services are often achieved through hardware devices, including hardware firewall, load balancer and VPN gateway. The advantage of hardware devices lies on their high forwarding capacities. But they also have the shortcomings of complicated and inflexible networking and weak virtualized functions. For instance, the number of virtual firewalls supported by the hardware firewall is limited, which can hardly provide independent value-added service of private firewalls for many tenants at the same time.

As NFV technology is growing and maturing, vFW (virtual Firewall), vLB (virtual Load Balancer) and vSSL/IPSEC GW (virtual SSL/IPSEC Gateway) are massively applied in the data center. First, these virtualized network services are operated based on a virtual machine. So their performance allocation and number can flexibly scale up and scale down on the basis of cloud computing technology. Second, network elements based on virtualization also guarantee that the resources of every tenant are monopolized and separated. Besides, services function chaining operated by SDN can achieve flexible value-added service orchestration, providing tenants with customizing services according to their needs.

Currently, manufacturers of many traditional hardware devices have introduced network service products with commercialized virtual services, such as H3C, F5 and Brocade. Parts of the products have been actually applied in the public and private cloud networks of the operators. However, because the whole market share of vFW, vLB and other virtualized network elements products is small, their advantage in cost is not obviously shown compared with hardware devices at this stage.

2.4 The Technology of NFV Flourishing and the Industry Chain Gradually Improving

Since the concept of NFV was first proposed in 2012, with the joint efforts of operators and vendors, it has been growing from an abstract framework to actually deployment. Currently, some operators have launched the deployment, testing and field tests related to NFV. Commercial deployment programs on NFV also have been continuously emerging. 2015 is considered as the first commercial year of NFV.

For the NFV industry, IT manufacturers are a significant power which cannot be ignored. The concept of NFV is revolutionary for traditional CT vendors. The traditional method for developing NEs based on dedicated hardware and private operating systems will be replaced by virtual network functions running on general-purpose hardware and general-purpose operating systems within the IT industry. IT manufacturers who actively join the NFV industry with their IT systems knowledge and experience become an important force in promoting the NFV industry. In the meantime, CT companies can not only get rich experience about IT systems from IT vendors, but also actively learn the IT industry development methods of open source and open development, which accelerates the development of NFV.

2.4.1   The Composition of the NFV Industry Chain

With the NFV industry's maturing gradually, NFV solutions have continuously been introduced. The manufacturers participating in NFV can be divided into the following categories.

1. System solutions provider

Generally such companies are large traditional CT vendors who have rich commercial experience of traditional CT network function. They can provide users with a whole set of system integrated solutions including hardware, NFV platforms and VNF software.

2. Platform provider

Such companies are typically IT vendors. They lack the experience of CT NEs, but they have accumulated rich experience over many years in IT systems. After the introduction of NFV, they apply the experience of IT platform systems to CT in certain transformations and enhancements and then supply the platforms to CT manufacturers as an NFV basic platforms. Such vendors are currently very common, such as Wind River and Red Hat, etc.

3. VNF provider

Such companies are often beginning companies in small scales. They have a little experience in the CT industry and make efforts to study in the field of the IT industry. However, compared to traditional CT giants, they lack enough power to achieve a bottom-to-up system. The NFV industry brings new opportunities to them. They join this industry by providing low-cost VNF software and provide users with complete NFV solutions relying on the software integration of the third-party integrators and platform vendors. The emergence of NFV remove the difficulties for entering the CT field, which means those small companies who was considered insignificant before by CT giants are also able to join in the competition.

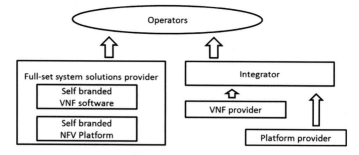

Fig. 18 NFV industry chain

The participation of such companies also brings chances to reduce costs for NFV industry. Currently the typical companies of this kind are Clearpath and Certusnet Info Technology Co., etc.

4. The integrators

Such firms tend to be relatively strong and have rich experience in the CT and IT fields. They are one of the companies in the categories above, but they can also provide integration services. They integrate hardware, NFV platform software, and VNF software from different vendors together to provide users with a complete solution. When deploying NFV, operators will face a question that who is responsible for interoperability problem introduced by hardware and software decoupling. In fact, generally, the answer is: the integrators. Integrators have rich integration experience in the NFV industry and they have a thorough and bottom-to-up understanding of the entire ecosystem. They can provide integration services for operators. Currently the typical companies of this type are HP, etc. the NFV supply chain is shown in Fig. 18.

It can be seen that with the gradual maturation of NFV, the number of manufacturers participating in NFV has increased gradually. In addition to the traditional CT giants, traditional IT vendors and emerging start-up companies have also joined NFV. The emerging of NFV has injected new blood into the CT industry and therefore the entire CT industry chain may be redefined.

2.4.2   Standardization and Open Source Communities of the NFV Industry Jointly Promote the Realization of NFV

The NFV industry is a typical example of a combination of CT and IT. Traditional CT VNF, set up in a virtualized environment built by IT systems, can provide users with open and flexible NFV solutions. In order to promote the development of the NFV industry, operators and vendors not only actively promote the practice of standards of NFV in standardization organizations of traditional CT

Fig. 19 Standardization organizations and open source organizations in NFV

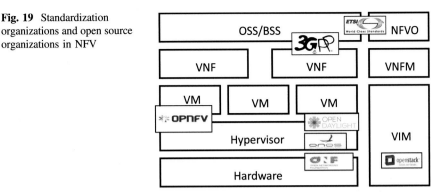

areas, but also actively learn from the experience of the IT industry to accelerate the realization of NFV relying on open source communities.

The main organizations for standardization relating to the development of the NFV Industrial Development are ETSI (European Telecommunications Standards Institute) and 3GPP (The Third Generation Partnership Project). Open source organizations mainly include OPNFV (Open Organization of Network Function Virtualization), OpenStack, ODL and ONOS, etc. wherein, ETSI NFV ISG founded NFV and formulated NFV architecture. 3GPP is responsible for the introduction of NFV in mobile networks and network management. OPNFV, OpenStack, ODL and ONOS are all open source organizations who are responsible for different software modules, which are shown in Fig. 19.

1. The development of relevant standardization organizations

ETSI NFV ISG and the Open Network Foundation are the founding organizations of NFV and SDN separately and they are also the main architecture constitutor and the driving force of this industry.

ETSI NFV ISG is the source of NFV. Since the NFV White Paper was first released in 2012, the organization has grown in large scale with 235 companies and 34 operators. Each conference has about 300 participants. ETSI is considered as the authoritative organization in the NFV field in the aspects of NFV architecture, demand and POC, etc.

3GPP is the standardization organization for mobile networks. Currently, 3GPP has finished standardization research involving business chains in the aspect of NFV in the SA2 working group. Now, it is studying the standardization of NFV related network management in the SA5 working group, mainly focusing on the development of specific standards of virtual network management architecture, OSS/BSS network interface requirements, collaborative processes of cloud management, network information modeling and new features and indicators after NFV. In June 2015, SA5 completed the study on the TR (Technical Report) stage and now, the study on the TS (technical standards) stage has been started. NFV network management standardization is expected to be completed by the end of 2016. 3GPP has also begun to consider making next-generation mobile network standards based on NFV/SDN technology and NFV platform layers will depend on the work progress of ETSI NFV ISG and other open source organizations.

2. **The development of relevant open source organizations**

Under Telcommunication 4.0 architecture, different open source organizations are responsible for different software modules, which laid the foundation for the platforms in Telcommunication 4.0. The main relevant organizations are the following:

OPNFV: OPNFV is an open source organization initiated by companies led by China Mobile, the American Telephone and Telegraph Company and other telecom operators. It is responsible for the output of integrated NFV platforms by integrating a variety of divergent open source software to provide carrier-grade NFV platforms and solve the interface consistency between the open source modules. In June 2015, the organization released its first version 'Arno'.

OPNFV project aims to provide a carrier-grade open source platform for NFV industry with high reliability, high performance and high availability. The developing contents of the open source project currently contain NFVI and VIM and will probably further expand in the future. OPNFV has received extensive attention from the NFV industry ever since it was founded. The project was announced to start in September 2014 and it has had 54 member companies so far. The first summit of OPNFV in November 2015, which has just been completed, had more than 700 participants, which made OPNFV the open source organization with the most participants at the first summit.

OPNFV is an important step towards open source for NFV. The traditional IT industry has been developing rapidly relying on several open source organizations, which the NFV industry must learn during its development. The foundation of OPNFV is undoubtedly a significant advance for NFV from concept and framework design to the realization of real deployment. The realization of NFV based on virtual environment requires multiple open-source software modules to be integrated, including OpenStack, KVM (kernel-based virtual machine) and OVS, etc. while these traditional IT open source organizations do not naturally integrated. The first tasks for OPNFV is to integrate these divert open source components together so that they can coordinate with each other.

In June 2015, OPNFV released its first version of Arno. This version, including OpenStack, ODL, OVS, DPDK, KVM and other open-source software, provides two types of installers (Fuel and Forman) in order to supply an open-source NFV platform for the NFV laboratory test and POC. The release of this version is a milestone for OPNFV. On the one hand, it means that OPNFV has operated formally as an open-source organization. On the other hand, it provides the basic open-source platform used for testing and developing of NFV.

The OPNFV plans to release the second version of Brahmaputra in February 2016. Compared to Arno, Brahmaputra is expected to provide several SDN controller integrations including ODL, ONOS and Open Contrail and will also support more installers. At the same time, the enhancement made by OPNFV, such as fault detection and policy management will also be gradually integrated into the release versions. OPNFV expects that the third version which will be released in the latter half 2016 will support the actual deployment of NFV.

Compared to other open-source organizations, another attractive feature of OPNFV is its community laboratory project, namely Pharos. Currently there are dozens of OPNFV laboratories around the world which have connection to OPNFV CI (continuous integration) to providing integrated testing environment for OPNFV. Also these laboratories will be opened to their party developers for the learning and practicing of the released results of OPNFV. Some laboratories provide purely physical devices so that users can log in the labs and use these physical devices to build OPNFV platforms for testing, while others provide OPNFV platforms which have been set up already so that users can log in the labs and experience services provided by these OPNFV platforms. Currently, a large number of projects in OPNFV use OPNFV laboratory resources to develop and test codes.

OPNFV integrates a plurality of upstream open-source codes. And it also actively develops appropriate codes to satisfy special requirements of NFV. These new introduced codes will be output back to the corresponding open-source organizations, using the power of OPNFV communities to promote the acceptance by the upstream organizations. On the other hand, these codes will also be integrated into the new released codes of OPNFV after their maturity. OPNFV use this method to make sure that the code will stay close with the upstream organizations and also to meet the special requirements of NFV promptly. Currently there are 45 projects which have been adopted in OPNFV already. These projects, integrating various requirements of NFV on resource management, high availability and performance acceleration and on the basis of the analysis of the demand scenes, developing codes and test cases gradually, input OPNFV CI to become part of the release versions of OPNFV.

OpenStack: OpenStack is an open-source project which was initiated by NASA and Rackspace and authorized by Apache license. OpenStack aims to develop an open-source cloud management platform, construct and manage public and private clouds and provide users with IaaS (Infrastructure as a Service) like Amazon EC2. OpenStack supports almost all types of cloud environments. The project goal is to provide a cloud management platform which is simple to implement, massively scalable, rich in resources and unified in standards. Wherein, each service provides rich API for integration.

The open source organization of OpenStack has been developing into an organization with 32,936 developers and more than 20 million lines of codes today from one with 573 developers and 55 million lines of codes in just three years since its establishment in September 2012. Through the year, it has become an IT open-source organization with worldwide attention. The development of OpenStack promotes hundreds of cloud computing companies to start and take off and the development of the entire cloud computing industry.

The rapid development of cloud computing provides a solid foundation for the realization of NFV. Because of its important role in cloud computing management, OpenStack has also been selected as the virtual resource management platform by the NFV industry. With the growing influence of NFV, the Power of the NFV industry is expanding within the open-source organization of OpenStack. Telco WG in OpenStack focuses on studying operators' enhancement requirements for OpenStack at the NFV scenarios. At the same time, there are a lot of NFV users among the OpenStack users group and they will propose users' requirements on the NFV scenarios.

OVS/DPDK/KVM: OVS/DPDK/KVM are open source organizations responsible for different network modules and components. OVS is responsible for the open source of virtual switches, KVM responsible for the open source of the server virtualization platforms and DPDK is responsible for the open source of the acceleration software for data plane.

2.5 Still a Long Way to Go: More Progress Should Be Made in NFV Technology

Carriers have great expectations for the future of NFV, hoping it could tackle the current pain points in operations. However, the carriers around the world have found that there is still a long way for NFV to be commercialized in large scale. Before the commercial deployment of NFV, the carriers should solve many problems, not only in techniques but also in operation and management. This section will mainly cover the technical challenges for NFV.

1. Unaccustomed to CT, IT needs to be optimized and adjusted for NFV

Cloud computing technology has been mature in IT industry. This technology is now widely applied in CT industry in order to provide NFV with a virtual and open operating environment. Nevertheless, cloud computing technology was not originating from CT industry, and there are several differences between the demands of the CT and IT industry. Therefore, when it is applied to the CT industry, it should be enhanced in some ways.

The most significant IT technology that has been applied to NFV is OpenStack. NFV plans to utilize OpenStack, an IT Cloud management software, as the management software of its own Telecom Cloud. However, compared with traditional Cloud Computing management, the VIM functional module in the context of NFV has some new requirements for OpenStack. These requirements can be summarized into the following categories:

(1) Resource Management

In the context of traditional Cloud Computing, OpenStack plays its role in managing virtual resources. There is no need for OpenStack to know how underlying hardware runs and how each of them connects with virtual resources, but just manages the virtual resources and provides them to network tenants. However in the context of NFV, OpenStack works as manager of the entire NFV infrastructure, which consists of traditional virtual resources as well as hardware resources, network resources and so on. In addition, in the context of NFV, the users of OpenStack—orchestrator and VNFM, have more accurate requirements for underlying resources. For example, the orchestrator requires OpenStack to know exactly on which hardware a certain virtual machines deployed on in order to satisfy some specific VNF requirements of affinity/inaffinity deployment; or VNFM requires that OpenStack could reserve underlying hardware resources so as to make sure a huge traffic at a given time could be digested by new network elements in time. Currently in OpenStack, projects like Congress are actively promoting the realization of resource management strategies.

(2) **Fault Detection**

Traditional IT Cloud Computing management components have limited function in detecting faults of virtual resources. For example, the module of OpenStack's internal monitor—Ceilometer, has limited monitoring data sources and most of the data sources it receives are of little value for monitoring in the NFV scenarios. NFV users usually have the following requirements for fault detection on underlying layer.

As for the timeliness of fault detection, the failover time of traditional telecom network elements is usually regulated within 50 ms, which means that telecom network elements should detect the fault and complete the active/standby failover within a short time of 50 ms. Only then can the telecom service avoid being influenced by the faults. In the context of NFV, we certainly should not simply depend on underlying NFVI to detect the fault and fix it within 50 ms. However, the sooner underlying faults are detected, the whole system is more likely to respond promptly, restore the system properly and make it back to its normal state as soon as possible.

As for the range of faults being detected, our current OpenStack mainly focuses on the faults of its own business, while in the context of NFV, OpenStack is required to do more than that. Users also require OpenStack to monitor virtual and physical facilities, report and respond immediately once it detects the fault.

As for the granularity of fault detection, the quantity of detection items is relatively small since Ceilometer arose late in the OpenStack project and has not yet completely mature by a long way. In the NFV context, however, users not only need to know the running condition of underlying facilities, but also expect to evaluate the underlying facilities and even forecast the faults through multiple operation data.

OpenStack's limitation in fault detection has drawn the attention of the OpenStack Community. Now, both the Ceilometer project and the emerging Monasca project are actively promoting the establishment of fault detection API.

(3) **High Availability**

Talking about IT and CT's requirements for the reliability of OpenStack, there is a joke of "cattle or pets". People of IT industry think that virtual resources are "cattle", which should be managed as a group but do not require special attention to each of them. When one virtual resource breaks down, we just need to start a new virtual resource to fill this vacancy. In the CT industry, however, every fault is likely to cause errors and even paralysis in the telecom business, because it is telecom network elements that run in virtual resources. Therefore, CT carriers must regard these virtual resources as their "pets" and lay special emphasis on each of them. Once a fault occurs, it should be detected and fixed immediately for the sake of operating business efficiently on the resources.

At present, what OpenStack has been dedicated to is basically guaranteeing itself with a high reliability. Nevertheless, there has been no mechanism ensuring highly reliable virtual resources. Even when safeguarding the reliability of its own business, the fault detection and system restoration will respectively take about 5 s. What's more, when the software guaranteeing high reliability breaks down and needs redeployment, all the OpenStack business based on the software will be completely destroyed and must be redeployed. In OpenStack, there has been no mechanism protecting the high reliability of virtual resources, which in fact depends entirely on the upper-level VNF reliability-safeguarding mechanism. This means that, while based on virtual resources without any reliability protection mechanism, the upper-level VNF has to ensure 99.999% of telecom reliability, which is definitely too demanding for VNF. This high demand for reliability, therefore, has largely raised the threshold for VNF manufacturers.

Currently, CT carriers and NFV manufacturers have paid attention to this OpenStack's reliability issue, and they are positively enhancing the realization of high reliability in OpenStack projects. The Senlin project, for example, aims at equipping OpenStack with cluster management business and thus satisfies the demand for the reliability of virtual resources in OpenStack. However, the traditional IT industry has set many obstacles for them. The major purpose of OpenStack, after all, is to establish a management platform for IT Cloud Computing rather than develop its high reliability. The OpenStack originating from the IT industry could never satisfy all kinds of NFV's demands for OpenStack. Indiscriminately copying from the OpenStack of IT industry may consequently bring challenges and risks related to availability, reliability as well as operation and maintenance management to carriers. With further understanding of NFV and richer experiences in deployment and practice, carriers have been increasingly more concerned with the demands for IT software like OpenStack at the carriers' level. In the Telco workgroup of OpenStack and OPNFV, an emerging open source project of NFV, carriers are jointly promoting the faster realization of demands at the carriers' level.

2. From the perspective of the industry, NFV technology needs further improvement

Since the concept of NFV was first brought out in 2012, the NFV industry has developed rapidly. Traditional CT manufacturers, IT manufacturers, new NFV software providers and IT equipment manufacturers all devoted themselves into the promotion and practice of the NFV industry, struggling for a place in this emerging and booming industry. Nevertheless, there remains a lot to be improved about this emerging industry. Through practice, carriers and manufacturers have been gradually identifying and solving new problems one after another.

(1) **Data plane acceleration schemes for virtual network elements needs unified interface**

Since telecom business demands high throughput and being real-time, the performance of virtual network elements on the data plane has always been a bottleneck for comprehensive deployment of NFV by means of the common IT platform and virtualization technology. The essential problem is that currently popular IA (Intel Architecture), large-core processor of common IT platforms and protocol stacks for universal operating systems are not able to efficiently deal with a great number of simple but concurrent data planes, i.e., message processing and forwarding.

To manage the performance of virtual network elements on a data plane based on a common IT platform, various manufacturers have carried out software and hardware solutions with respect to different network functions (encryption/decryption, content filtering, virus protection, audio/video coding/decoding, third–seventh layer exchange and so on).

A representative software solution is Intel's DPDK, which aims at solving the problem of low efficiency of the Linux network protocol stack. DPDK makes full use of the capabilities of the x86 processor by optimizing the user state packet processing software, which makes it possible to implement efficient packet processing on the COTS server of IA.

Hardware solutions include local specialized accelerator chips that can shunt universal CPU packet processing work (such as intelligent Ethernet adaptor), SoC (System on Chip) hardware platform optimized by collocation of big and small cores (such as ARM architecture server), and specialized separate-type accelerating resource pools. There are diversified forms of hardware accelerator chips, including universal chips, specialized chips and dynamic programmable chips, such as FPGA (Field Programmable Gate Array) and DSP (Digital Signal Processor).

However, due to the lack of a unified interface and protocol standard for these solutions, VNF software on the data plane, which uses these acceleration technologies, will form a binding relationship with the corresponding system and hardware architecture/chips. This has made it difficult to decouple NFV functions from hardware devices. Therefore, researching on the universal acceleration architecture for VFN on the data plane of the mobile core network, as well as providing acceleration resource utilization/management interactive interface and protocol independent of underlying resources, will advance the comprehensive deployment of NFV in the mobile core network.

(2) **Compatibility of various kinds of software in NFV should be tested**

Since software and hardware have been decoupled, there arise various kinds of software for NFV to choose. On the NFV platform, Cloud management software can be OpenStack or Cloud Stack; SDN controller can be ODL controller, ONOS controller or Open Contrail controller; there are even more choices for the host operating system, namely Ubuntu, CentOS[5] and so on; and Hypervisor can be

---

[5]Editor's note: Both Ubantu and CentOS are distributions of Linux Operating System.

KVM, VMware and other commercial products. It is urge for NFV industry to deal with the issue of interactive operation and software compatibility. Generally, the software that manufacturers can offer is simply a verified integrated-compatible version, but it still needs to be examined whether the software replacement will affect normal operation on the whole. In order to ensure the flexible deployment and openness of NFV, software compatibility testing must be accomplished in the future.

(3) **The NFV platform should be decoupled with the VNF software**

From the perspective of NFV architecture, NFV platform can be decoupled from VNF software. In deployment practice, however, some manufacturers are often more inclined to provide products that couple VNF and the platform for the purpose of strengthening their dominance on the market. Manufacturers indicate that with products coupling VNF software and NFV platform produced by the same manufacturer, reliability and high performance of the whole system can be guaranteed, which will be difficult if VNF and the platform are from different manufacturers.

However, if we couple VNF and the NFV platform from the same manufacturer, the deployment of communication network then will become a difficult thing. In the communication network, it is impossible to use the whole scheme for NFV from only one manufacturer. Different types of VNF may come from a variety of manufacturers. Binding VNF and the NFV platform together means that every time procuring a new VNF should be coupled with a new NFV platform. NFV platforms from different manufacturers are unable to share hardware resources or dispatch software resources to each other, which will largely reduce the flexibility and resource utilization rate of the network. What's more, we can easily say that nothing really changes compared with the traditional procurement pattern, in which hardware and software were closely coupled. Though hardware is decoupled from software now, manufacturers can still raise the price through binding VNF and the NFV platform. Carriers could have just purchased one set of NFV platform software for various data centers to use, but now they have to pay for a number of NFV platforms.

Therefore, in the practice of NFV deployment, it is extremely significant to decouple VNF and the NFV platform. On the other hand, it is also essential for carriers if they aim to realize NFV and take full advantage of this new network architecture.

(4) **There is a big gap in the realizations of NFV orchestrator**

The NFV Orchestrator, as the brain of NFV, connects NFV with the traditional communication network management system and it is a crucial part of NFV deployment. There have already been a number of manufacturers working on the development of NFV Orchestrator. Nevertheless, current NFV Orchestrator integrate unsatisfactorily with the traditional network management system. They are suitable for individual deployment and demonstration, but there may be some problem with practical deployment in the traditional communication network. The traditional communication network management system is usually designed and developed by carriers themselves, who usually have their own special requirements. Thus, NFV Orchestrator running in a carrier's network should adapt to the carrier's network manager. In the future, carriers should be actively involved in the design work of NFV Orchestrator, and pay much more attention to the Orchestrator's adaptation to traditional communication network managers.

(5) The degree of maturity of different VNF products varies

At present, VNF manufacturers in the NFV industry have laid much emphasis on the virtualization of the core network and have developed a variety of core network virtual products. However, most of these products are still disposed according to the classification of the traditional CT network equipment's functions. Individual VNF often occupies 8 or even more CPU cores, which is not ideal as an NFV product. Future network virtual products need more granular classification in functions. Making the best use of virtualization technology, the flexible, fast and highly reliable deployment and maintenance of the core network can be realized through small virtual network elements in various degrees of granularity.

There are also other VNFs besides core network virtual products, such as virtualization access network elements. Since the deployment scenarios of carriers differ a lot from each other and are still developing, the degree of maturity of these VNF products is relatively low. Traditional CT giants are often unmotivated in developing these VNF products. Though traditional accessing equipment has been widely deployed in the communication network, the way of deployment and functions of the new VNF remain unclear. Therefore, traditional CT manufacturers have invested in this area with much discretion. By contrast, emerging VNF providers, especially some startups, become increasingly more interested in network elements virtualization of the access network. Their scales are often relatively small, uncompetitive in core network virtualization with large CT manufacturers, so access network virtualization has been a crucial chance for their development. However, these emerging manufacturers have limited experience in the communication network and do not fully understand the carriers' requirement on network operation, maintenance and management. As a result, it is not that easy for carriers to promote virtualization in the access network.

(6) Standard interface between NFV and SDN is confronted with challenges

NFV lacks the capacity of interworking with SDN standards. Connections between the two independent architectures, SDN and NFV, are concentrated in the data center and Cloud site scenarios. The critical operation that chains them together is the operation chain. Communications between SDN and NFV should have been based on the open, standard northbound interface. However, OpenStack's current operation chain programs (such as GBP—Group-based Policies) have no complete functions, and it is impossible to arrange and drain SDN to individual VNF. Therefore, SDN manufacturers usually adopt a private customized SDN API interface to interwork with NFV, resulting in a binding relationship between SDN and NFV products.

In addition, NFV has had a profound impact on the entire communication industrial chain, including not only technology and virtualization products but also management, operation and maintenance of the entire communication network. Chapter 5 of this book will then focus on NFV's profound impact on the communication industry.

3   Telecommunication 4.0 New Network Architecture Based on NFV or SDN

Telecommunication 4.0 is an integration of IT and CT. It will completely change the network in a subversive way and forms a new network architecture, which is mainly reflected in the following ways:

Firstly, the physical architecture of the telecommunication network in the future might be composed of standardized data center nodes, and its logical structure will be hardware and software decoupling. The software networks are deployed in a unified cloud infrastructure and the infrastructure resources can be fully shared.

Secondly, the new network will be capable of managing, orchestrating and scheduling the whole network in a uniform way, implementing a whole life cycle management of the virtual network, flexibly allocating and adjusting the network connection of the data center and across data center.

Thirdly, the new network will form an opening architecture, enabling it to open wider and provide customer service as soon as possible. A universal, open and adjustable API will play a key role.

1. Completely Change Network Architecture and Form a New Network Based on Data Centers

The telecommunication network is made up of millions of telecommunication machine rooms physically. Over the past a few decades of fast development, the telecommunication network almost doubles in scale each year, and the quantity of equipments and complexity of network also saw a drastic increase and the network pattern became more complex, as Fig. 20 shows. Over the past few decades of development, the traditional telecommunication machine rooms became more

Fig. 20 The telecommunication room number in the whole network

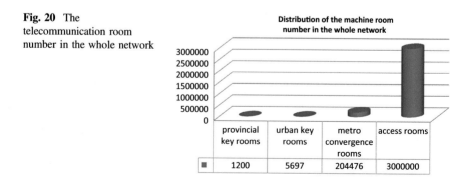

	provincial key rooms	urban key rooms	metro convergence rooms	access rooms
■	1200	5697	204476	3000000

complex as each room differs from the others in equipments, cable size and environment, thus causing a series of challenges like the difficulty in expanding network capacity, large load on network maintenance and operation, complicated management, high costs of network deployment and so on.

By the end of 2015, the three telecommunication operators in China owned over 1200 provincial backbone core machine rooms, 5697 metropolitan core machine rooms, more than 204,476 metropolitan aggregation machines rooms and more than 3 million access machine rooms.

With the development of new technologies like NFV and SDN, we are glad to find some opportunities to solve those challenges. We can redesign the telecommunication network infrastructure just like using Lego bricks and by doing so we can completely get rid of the difficulties.

Under the Telecommunication 4.0 network design, the whole network will be divided into two parts: the infrastructure and network application.

As the shared platform of the whole network, the infrastructure is based on a component or TIC with a reproducible, extensible NFV and a high SDN-intensive data center network node. TIC, the basic cell of the Telecommunication 4.0 network architecture, is the basic unit for reestablishment of the telecommunication network. By carrying different network functions, TIC components can be deployed in the network access edge, provincial center network, regional center network and national center network in accordance with the needs of network deployment.

The network application, responsible for the basic functions of the new network, could be deployed on infrastructure flexibly in accordance with service needs.

Figure 21 is an example of the TIC components of the Telecommunication 4.0 network. We can see in this picture that the new network application is software-based, the hardware and software have been completely decoupled, and the updating of functional software is not forcibly attached to hardware. As a result, network

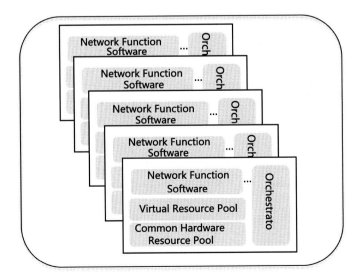

Fig. 21 The TIC components of the Telecommunication 4.0 network

application could be flexibly deployed in different TIC in accordance with service needs. TIC is the key to our Telecommunication 4.0 research, and we can take it as a universal container for the future network. The container has the following features:

First, complete decoupling; under which the virtual operation environment and network function application software are fully decoupled and are able to be allocated freely and flexibly, so they are not limited to a binding relationship between connectors.

Second, based on universal IT equipment, the hardware resources are unified and can be shared flexibly. In addition, upper application could also be removed flexibly.

Third, shared resources; by providing a unified virtual operation environment through the universal hardware platform for network application, hardware resources could be used universally and virtual resources will be able to be shared.

Fourth, a programmable network; by flexibly programming calculation resources, network resources and storage resources, network will be able to be adjusted dynamically and to rapidly respond to the demands of the market.

As Fig. 22 shows, Telecommunication 4.0 network architecture is just made up of TIC components. In accordance with the needs of service and deployment, it has formed the network edge access node, provincial and regional center node and national center node. The service system could be deployed flexibly in accordance with its own features. For example, BRAS, SAE-GW(separate area evolution

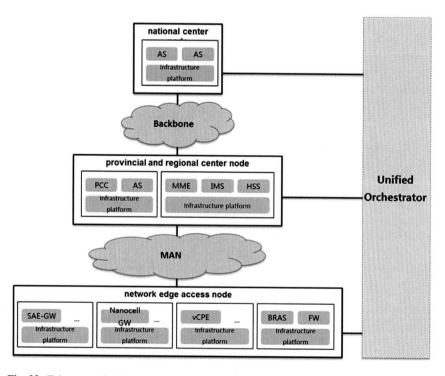

Fig. 22 Telecommunication 4.0 network architecture

gateway) can be deployed to the network access edge node, IMS, MME(mobile management entity) can be deployed to the provincial center node or regional center node, while the application platform AS(application server) can be deployed to the regional center node or national center node.

Telecommunication operators in the world all think the same in reestablishing a new network infrastructure. The US telephone and telegram corporation realized the weakness of the current network in 2013 and came up with AIC in its Domain 2.0 plan (software network and virtual network function plan), in which it emphasized using the cloud framework of AIC establishing a shared and standard network cloud platform and creating a shared cloud network environment. Tim Harden, president of the US telephone and telegram corporation's supply chains, pointed out in the announcement of the Domain 2.0 plan: we are now thoroughly improving our expansion and operation methods so as to manage our service more easily, just like managing data in the data center."

2. Comprehensive Program and Management—The Key Function of Telecommunication 4.0

In terms of the design of the telecommunication network system, system featuring scattered data and centralized control has been the consensus of the telecommunication sector. However, as for how to establish such a system, the centralization degree of the system changes with the difference in the function and range of management and control.

SDN technology helps us control, transmit and separate networks, while NFV technology gives us another hand. The integration of the two technologies helps us get a whole picture of the network and improve it comprehensively. In Telecommunication 4.0, comprehensive management and programing of the whole network is the key to comprehensive optimization of the network, as it perceives the performance of the network node better, which ensures a flexible resource adjustment and fast launch of new service.

Google has provided us with a very good example. *Andromeda*, a virtual network system Google developed independently, doesn't use the Autonomous System, which was widely used in the Internet, instead, it uses a SDN controller-based network model subjected to centralized control, which covers an end-to-end view of the whole network. In the end-to-end view, all the network access information and equipment capacity information are stored in an application program. So, based on source address and destination address, the SDN controller can calculate the route between them, decide different network routes for different traffic data types and ensure a quick response to the changing network environment. With this model, Google could achieve automation by controlling and programing the whole network through a single allocated node.

(1) Different Content Arrangements between NFV and SDN

In the framework of Telecommunication 4.0, as for its objects, layout management capabilities that we are facing can be divided into two categories: (1) network services and VNF deployments, and (2) network routing configurations.

VNF network service and deployment adopts NFV orchestration capabilities, and NFV Orchestrator calls VNFM and VIM on NFVI resource allocation and scheduling to implement the network services, automated deployments of VNF and on-demand capacity expansion and contraction.

Networks and routing configurations are in the internal data center. It uses the SDN orchestration capabilities and sends control information to hardware switches, virtual switches, virtual routers, virtual firewalls, virtual load balancers and other devices in the SDN controller data center through the SDN controller so as to complete the above NE information configurations. WAN has an SDN controller controlling the bearer networks by way of devices and transmission equipment to achieve flexible routing settings. In some scenarios, according to the needs of deployment, a grading SDN controller may be set. We can control the entire network router through controlling two stages, which are the local controller and the super controller.

From the single scenarios, the deployment cases that use NFV or SDN technology are successful, largely solving problems encountered in current networks. But comprehensively, the deployments in each scenario are still isolated. From the perspective of network automation, it still doesn't meet the requirements of Telecommunication 4.0. So, in the era of Telecommunication 4.0, what we need to do is to have a deep integration with the NFV's and SDN's arrangements, forming a unified global choreography management capability.

(2) A Unified Global Management Is the Key to Achieve Global Optimization

We can understand it in this way: NFV's choreography capacity is equivalent to a person's nervous system, whose function is equivalent to a person's muscles, while SDN helps us to build a person's blood system. All of these systems are summarized in human's brains, accepting the brain's unified control. The brain can be regarded as a global management system. This shows the importance of the overall choreography management. If we didn't arrange such brain control, we couldn't create a dynamic and flexible network.

The overall arrangement abilities for Telecommunication 4.0 need NFV's arrangement abilities and SDN's unified traffic scheduling capabilities. On the one hand, a unified arrangement ability coordinates with VNFM in NFV to accomplish the network's choreography arrangements, network services, VNF's life cycle managements and overall resource managements. On the other hand, it coordinates with the SDN controller to realize the interconnection, and establish business demand chains as needed, forming a virtual network, so as to accomplish resource exchange visit and integration. With this overall choreography arrangement capability, we are able to allocate integrative all network resources, including routers, network application software, data center configurations, and so on. And we cooperate with these different network resources to achieve a shared, dynamic, programmable network.

In the era of Telecommunication 4.0, the overall choreography management ability makes our work more simplified and convenient. We put up with the needs for the deployment of the new network and new network functions on the unified choreography arrangement system while the system will be linked to the need to allocate the demanded resources, including computing resources, storage resources in the data center computer room, and transmission resources across the entire network. The overall arrangement will enable the network operator and maintenance personnel to have a clearer network view for the new on-line networking features. The management of network resources will have unified planning, which is no longer divided into systems and filed systematically. The resources of the entire network system will be fully utilized.

3. The Endogenous Open Ability and the Universal API

The rapid development of the internet gives telecommunication business innovations new challenges. How to find its right position in this "drama" of future mobile internet, and play its role is the key for the Telecom operators in the era of Telecommunication 4.0 to whether it can have a successful transformation or not.

In the process of seeking win-win cooperation commonly between the telecommunications networks and the internet, opened network ability has always been a hot topic. Open the basic capacities of the telecommunications network to the third parties and establish a channel of interconnection between networks and business, which not only can promote the third-party mobile internet applications innovations continuously to offer users richer new services, but also make the telecom network optimize its network resources according to different business needs.

Open communication is not a new product in the era of Telecommunication 4.0. Before Telecommunication 4.0, we were familiar with opening to the network capability. Around 2011, all types of platforms emerged. All these open platforms had similar design concepts, which made the network capacity into the API package and can be called through for the third-party applications with the upper open platform in the networks.

In the era of Telecommunication 4.0, opening the network ability will no longer be realized in the form of an opened platform, with NFV and SDN technologies used for reference, networks form an endogenous open framework via a common API, its open ability will be further enhanced, realizing the integration of network ability according to demands. Endogenous network ability open frameworks accomplish their basic infrastructure capacities and value-added services opening to the third-party mobile network by means of making effective and proper abstracts for network functions and shielding the underlying network technical details, enabling telecommunications networks operators to transform from a closed system to an open and inclusive interaction networking system.

NFV standard framework consists of endogenous open frameworks, the network ability can open to the outside world by means of standard opened API. The hardware resources of NFVI, including computing hardware, storage hardware, network hardware resources, can be used by virtual resource groups and VNF by virtue of standard interface opening. And the virtual resources of NFVI, including virtual computing, virtual storage and virtual network resources, can be called by VNF by means of standard interfaces. The arrangement abilities of NFVI are open

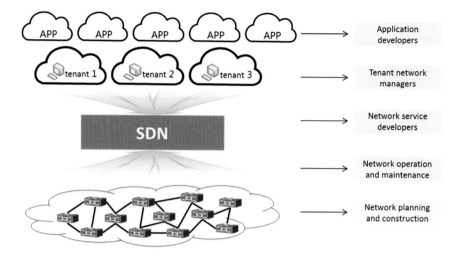

Fig. 23 SDN architecture diagram

to application parties to achieve fast on-line business by opening interfaces of Service Orchestrator.

In the process of applications of SDN data centers (Fig. 23), the SDN controller opens the network capacities to the application layer by means of northbound interfaces. Neither do the application layers need to know or understand the complex structures of the network. Based on opened network frameworks and capabilities, we can achieve a variety of innovative network analysis applications, arrangements and scheduling to meet the demands for rapid opening of the business and flexible deployments.

Universal API is the key to achieving endogenous opening frameworks. It requires that the industrial community adopts a variety of methods to carry forward, such as standard, open source, certification and so on. Only in this way, can we truly meet the fundamental requirements of cross-platform and universal.

4. Component Standardization and Integrated Arrangement Networks

The targeted framework of Telecommunication 4.0 is to combine NFV and SDN technologies together with depth to achieve resource sharing, flexible scheduling and dynamic programming of the entire network. The Communications 4.0 network will be a network with standard components and integrated arrangements. Among them:

Component standardization refers to the use of unified planning, duplicable, sharable, scalable TIC components to replace new types of carrier-class data centers formed by traditional telecommunications rooms, and share physical resources and transmission resources with the traditional data centers.

The integrated arrangement means to achieve global optimization by means of unified network scheduling and dispatch management functions.

Chapter 4
The Dawn of Telecommunication 4.0

Before the Telecommunication 4.0 era, CT industry was relatively isolated and rarely overlapped with IT industry. CT and IT were two ecosystems evolving in parallel. With the advent of the Telecommunication 4.0 era, deep integration of CT and IT industries will breed a new industrial ecosystem, and become a huge industrial revolution. The reconstruction of industrial ecosystem will result in a brand-new industrial upgrading, all participants will do their best to take an advantageous position in the process of industry reconstruction. But during the process of revolution, what we need is an open mind and win-win cooperation, rather than competition.

For CT industry, the convergence and close interaction between CT and IT is not only an opportunity but also a significant challenge. Currently, China is in a critical period of industrial transformation. The CT industrial integration and successful transformation brought by the Telecommunication 4.0 will greatly stimulate the demand of national information consumption, "Internet+" and public innovation as well as unleash enormous industrial momentum. CT industry should embrace the future, keep moving forward and drive innovations with an open mind to lay solid foundation for the development of the national economy, and transform China from a large CT country to a leading CT country.

1 Game Brought by Telecommunication 4.0—Deep Integration of CT and IT, Emerging New Leaders

1.1 Problems Behind Rapid Development of Traditional CT Ecosystem

Due to the complicated system design and relatively closed network and business systems, it is difficult to enter CT ecosystem. Only a small number of companies are

© Springer Nature Singapore Pte Ltd. 2018
Z. Li, *Telecommunication 4.0*, DOI 10.1007/978-981-10-6301-5_4

capable to get in, fewer can provide a whole solution, and even less is able to run an end-to-end network.

Even since a long time ago, the CAPEX of telecommunication systems stays high compared to Internet OTT service. Because of the closed supply chain, lack of openness and agility, telecommunication network is not able to respond quickly to the ever changing customer requirements.

1. High CAPEX due to the closed industry ecosystem

In traditional CT industry, the operators are in the dominant position. Equipment vendors are responsible for offering total solution and guarantee the system quality, while upstream chip and device vendors will provide dedicated and customized carrier-grade hardware to the equipment vendors.

Due to the high-performance and high-reliability requirements, design and R&D costs of telecommunication equipment is higher than that of the general IT equipment, from both software and hardware perspective. Besides, since only a few manufacturers is able to build these equipment, the lack of competition results in high procurement cost. With the rise of domestic manufacturers and improvement of public bidding process, the procurement cost has declined. The overall investment and customer service cost is still higher than that of general IT equipment.

The communication networks are strictly required to offer high reliability and long-term security service. Load of the operating equipment should stay low under normal traffic model. Even if in the event of extreme case such as the Spring Festival, the load generally stays within 70%, which leads to a waste of resources.

Because of the manufacturers' proprietary software and hardware, it is not possible to share resources cross different platforms. Moreover, once deployed, the old hardware platforms might not be able to support newly installed software applications. In traditional network operation, it's not a rare case when an entire hardware equipment has to be replaced because it can't be upgraded to support new functions, which is also a waste of investment.

2. Limited network capability exposure leads to limited service

Traditional communication networks mainly offer basic voice and narrow-band data service. With the improvement of network capacity, HD audio and video service and broadband data service are then available in 4G era. However, services provided by the telecommunication networks are still limited. Previous killer applications, such as 2G voice, short message and Monternet have almost been forgotten, while new voice and data services, such as VoLTE and RCS, is still under development.

Entering into the Intelligent Network (IN) era, CT industry has come up with the "open network" slogan. IN services and value added services were once very popular, but since they were built on top of a closed systems, service expansion is not flexible enough. Each time of the service requirement changed, the service logics should be redesigned again. And this redesign is very complicated for a non-professional staff. In the IP era, people invented IMS, the IP multi-media sub-system, to try to add more type of voice services and use EPC, the packed core network to open the possibility to add data services. However, IMS can only solve service serial trigger problem, not service expansion, while EPC can only solve the load guarantee problem in a multi-service multi-users scenario, not the service expansion as well. In the IP era, although the network is more sophisticated, almost no telecommunication services can be remembered by customers.

Reality is always worse than what was wished for. Finally we realized that the network technology itself cannot solve the network capability exposure problem. Piling service platforms above networks still cannot add to diversity of telecommunication services. The root cause for limited network services is the limited network capability exposure of the entire communication network.

Even in the Telecommunication 4.0 era, we are still facing limited resources and shortage of talents in new services and applications, we should apply core competence to extract core capability of communication networks, bridge gaps between networks and service platforms and stimulate developer communities in telecommunication innovations.

3. Lack of agility due to communication service's long launch time

Traditionally, communication service should go through 5 key procedures from planning to launching, as shown in Fig. 4.1.

First, operators make technical specifications according to the international standards or customized requirements. Equipment vendors develop products accordingly. Then operators will conduct laboratory test and field trial to verify the products. After several rounds of verifications to confirm that the products are mature and conforming to the specifications, centralized procurement will be carried out. Finally the equipment will be deployed. The network operation department will then take over to ensure the network is in stable operation.

These 5 step processes (also called water-fall process) usually take years to finish. According to the actual network operation experience, it takes a year and a half from publishing specifications to stable online operation. From the network deployment perspective, the significant lag of this 5 step process cannot keep the pace with the highly developed information requirements in the Internet era (Fig. 4.2).

It is also very difficult to upgrade network functions and integrate new services into already complicated networks. Operators usually lag behind the Internet in terms of the collection of customers' requirements. Not to mention the time for transforming customers' demands to technical requirements, developing telecom specifications and driving vendors' R&D. Communication vendors usually take half a year to develop new products based on the specifications. By the time the products come online, chances are that this round of service requirements has already been outdated, or the customers have given up and resorted to the Internet solutions.

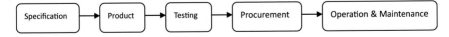

Fig. 4.1 Launching process of communication service products

Fig. 4.2 Launching process for Internet services

Basically, two barriers prevented communication products from entering the market. On one hand, operators and equipment suppliers are not closely connected in terms of the requirements and vendors' R&D. In other words, operators lack the ability to develop products independently and cannot transform customers' demands to products agilely. Thus the products always substantially lag behind requirements. On the other hand, the insufficient innovative thinking and creativity cannot satisfy the demand of the open "Internet+" era.

In the Telecommunication 4.0 era, the lengthy processes have to be changed. We should learn from the agile R&D processes and innovative thinking of IT industry. We should implement a close-loop, fast iterative and short process, while exposing network capability to seize the opportunity for market development.

1.2   Reconstruction of the CT Value Chain: Mutual Penetration Between IT and CT Giants

The traditional CT manufacturing chain mainly includes: the equipment suppliers use the chips and platform tools, buy and compile software, assemble products with black-box hardware platform and deliver them to customers (such as operators) for network deployment, as shown in Fig. 4.3. Generally speaking, customers and operators can only define specifications for interfaces between equipment and cannot decouple the software inside the equipment.

• Telecommunication 4.0 extended the whole industry value chain. By introducing IT general hardware, the hardware and software can be decoupled and IT can go deep into each link of the industry value chain. The black box hardware is first replaced by the commodity hardware with software functions, and the software function is further decoupled by introducing cloud computing and virtualized software. The core players of the new industry value chain are composed of manufacturers of commodity hardware, chip makers, software integrators, virtualization software providers, network software providers and cloud management system providers, as shown in Fig. 4.4. The importance of the software integration is amplified, including integration of cloud platform

Fig. 4.3 Manufacturing chain of traditional CT

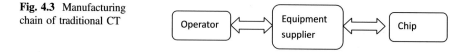

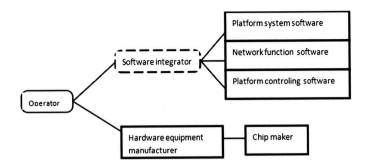

Fig. 4.4 Telecommunication 4.0 industrial manufacturing chain

software (from the virtualized software providers), the network function software (from network function software providers) and the cloud control software (from the cloud management system providers).

- From the development of Telecommunication 4.0 point of view, the original CT industry manufacturing chain will be broken by new IT partners (such as virtual platform providers, Red Hat, VMware, IT hardware equipment providers, HP and Lenovo). The traditional CT system vendors may switch to network function software providers or virtualization platform providers. In addition cloud management platform and SDN controller providers, as well as competitive system integrators are needed to deliver assembled products. It is an opportunity for both CT and IT, or even operators to rethink and reshape their business. This will definitely bring a new industrial value chain and trigger a new industrial cooperation model.

The Telecommunication 4.0 era imposes new requirements to both CT and IT market in terms of industrial reformation. In order to penetrate in the CT industry, almost all the big IT companies including HP, Intel and Red Hat have established a dedicated NFV/SDN department. While traditional CT companies, such as Ericsson, are trying to invest in IT companies to improve their competence. Some Chinese companies like Huawei have decided to step into both CT and IT to provide integrated solutions. In addition, all companies attempt to adopt open source culture in IT and the traditional standards practice in CT, to accelerate joint development and enable interoperability tests.

First of all, IT manufacturers are the active driver of this revolution. With the transformation of the manufacturing chain in Telecommunication 4.0, IT enterprises are seeking to enter into the new CT market, in the following three paths:

- IT chips are the basic components of Telecommunication 4.0. As a result, chip manufacturers are constantly improving their products to satisfy carrier-grade requirement to enter the telecommunications market. Their capability in general hardware and software is gradually upgraded to adapt to the CT industry. General processing chips has been used as part of traditional communication equipment, which proves the capability of these chips can meet the requirements

of telecommunication market. With the introduction of the software acceleration, the general IT chips can fully satisfy the carrier-grade performance
requirements. By providing the solution of general hardware server, LAN and
the accelerator card, x86 chip providers like Intel can naturally step into the CT
industry.

- The platform software is the core component of Telecommunication 4.0. As a
  result, platform manufacturers are launching new solutions to meet telecommunication demand. The traditional market size is large in terms of operating
  systems, virtualization and cloud management software. With the advent of
  Telecommunication 4.0 era, the communication market has a faster iteration
  speed and a larger network scale. Therefore, IT manufacturers are also des-
  perately to enter into the CT industry to provide software products that meet
  carrier-grade requirements.
- Software integration has become the "adhesive" of Telecommunication 4.0
  operation, attracting IT manufacturers transforming into system integrators.
  Some IT companies develop their own Telecommunication 4.0 ecosystem by
  procuring communication software solution companies and co-working with
  other software companies. In the meanwhile, they actively seek cooperation
  with other companies and drive interoperability test to equip themselves with
  system integration capability or merge their solutions into other companies'
  system integration solutions.

Secondly, CT manufacturers must reform themselves to face pressure from
industrial revolution and product transformation. CT manufacturers mainly bring in
IT capability in the following two ways:

- First, improve IT capability and directly provide new products. By establishing
  IT departments, CT manufacturers plan to enter the enterprise market to produce
  IT commodity hardware and develop IT platform software (including OS, virtualization and cloud management platforms). Companies like Huawei and ZTE
  strive to develop their own IT product lines and try to improve competence by
  combining strength with CT products.
- Second, introduce cooperation modes. CT giants attempt to gain IT competence
  in the emerging market either by investing in or cooperating with IT companies,
  such as Ericsson's alliance with the Open Source cloud service platform. This is
  similar to the methodology of building strategic cooperation with startups and IT
  giants to establish its own ecosystem, with which Alcatel-Lucent is actively
  building up its own NFV/SDN ecosystem.

Traditional CT leading players have rich network operation experience. In the
Telecommunication 4.0 era, they keep innovation and transformation in order to
keep their dominant position. In the NFV/SDN deployment, these companies turn
into system integrators and continue to provide services to operators. By
self-transformation, they are now capable to provide commodity hardware and IT

platform software, and thus can offer a whole set of products from hardware to software platform and to VNF. Otherwise, they could integrate with other vendors' hardware or software products and deliver the whole solution to operators. For operators, these leading CT vendors have rich experience and clear understanding of operators' requirements. They often cater to the operation mode of operators in terms of technical support and after-sale services. As a result, they can offer operators the most suitable products.

1.3   Integration of IT and CT Is Still Under Exploration

During the mutual penetration between IT and CT, they will also face many difficulties, because the business model, market competition or even terminology of the two are not well understood by each other.

CT vendors are not familiar with the business pattern in IT industries, such as the way of participating into Open Source community and services assurance by continuous software iteration. Upon entering the Open Source community, CT vendors are not used to the new game by using previous standardization processes. CT vendors want to improve their ranking and influence Open Source community, such that to take an advantageous position in the integration solutions of Telecommunication 4.0. This requires companies to make a series of reforms and institutional adjustments to attract a large number of Open Source software developers and experts to influence the community. IT industries embrace a win-win market, while traditional CT competition market hardly needs cooperation. Therefore, in the Telecommunication 4.0 era, CT needs a more open attitude to constitute a broader ecosystem and take an advantageous position in the final market competition.

As for IT manufacturers, they are even less familiar with CT market competition. Since they can hardly understand the carrier-grade high reliability, high performance and timely fault handling requirement, they just reuse enterprise-level solutions to operators directly. Therefore, their products can hardly meet the requirements of the CT industry in the initial verification. Besides, as new entrants of the CT value chain, IT companies need a relatively longer running-in time before being recognized and accepted by the CT industries.

IT manufacturers are also actively seeking changes. They are trying to compensate the traditional enterprise solutions shortages by improving the reliability of software products and integrate with other CT solutions. At the same time, IT chip makers are also aware of the capability shortage of general processors. So they are also developing acceleration techniques in order to improve the whole chip solution.

In conclusion, the IT and CT industries are gradually adapting to each other. A new industrial cooperation model will be formed in the near future.

1.4   Large-Scale Business Combination and Acquisition for IT and CT

There are three major types of business combination and acquisition of in Telecommunication 4.0: (1) IT manufacturers merge with CT software or startups and then enter into the CT market; (2) CT manufacturers invest into or acquire IT companies to keep their competence in the CT industry; CT manufacturers merge with each other to keep their own competence, as shown in Fig. 4.5.

On the 26th May 2015, HP acquired an SDN company—ConteXtream, which is based in California. Before that, ConteXtream's advanced technologies had already been deployed in many large operators' networks. This acquisition, together with previous acquisitions with many other software startups in the CT field, has turned HP into an ODL-based and carrier-grade SDN supplier.

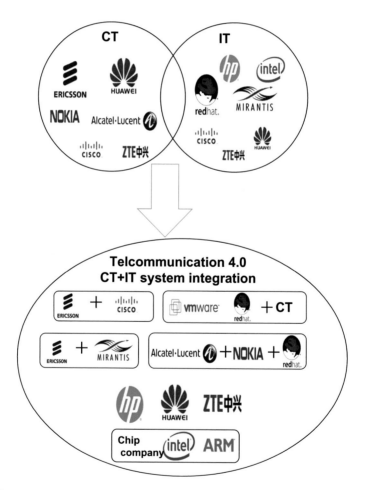

Fig. 4.5 Industrial Integration

In February 2015, Brocade Communications announced to acquire SteelApple series products from Riverbed. Only a month later, it again announced to acquire Connectem, a private company in EPC. The intention of these two acquisitions is to improve Brocade Communications' competence in SDN/NFV. After interviewed with Andrew Coward, vice president of strategy in Brocade Communications, he said the purpose of the acquisition is to complement their traditional products. Brocade Communications will focus on cloud computing and SDN technology to make the network become more efficient, more flexible and more cost effective. This acquisition will help them to take advantage of the existing basic equipment and hardware. By providing a packet gateway from Connectem to process and integrate data packets in different locations, each packet will have to be processed from just one time, which will greatly improve the processing efficiency. By this acquisition, integration between CT and IT in Brocade Communications will bring more effective support to the new industrial revolution. These are the two typical cases where IT manufacturers entered into the CT market by acquiring CT software startups.

On the 24th August, 2015, Intel and Ericsson announced to jointly invest $ 100 million in Mirantis to promote the application of OpenStack in enterprises. Established in California of US in 2009, Mirantis is an emerging cloud computing company in OpenStack. Judging from the current industry situation, although OpenStack has been widely applied among enterprises, for IT managers, OpenStack is still definitely too complicated. When enterprises started to use cloud services to manage business, a relatively large part of the business still need traditional IT system, whose operation hadn't taken the cloud services into consideration. As a result, in the transition stage, business operation will become unsettled. The investment of Intel and Ericsson in Mirantis aims at utilizing the cloud computing technology from Mirantis to simplify OpenStack to meet the cloud computing demand of the enterprises. In addition, they also hope to improve OpenStack's infrastructure management, elasticity and performance management of container and stacks. Meanwhile, Ericsson, the communication leading company, also wants to improve its own IT competence by investing in IT startups so as to maintain its technical leading role in the Telecommunication 4.0 era.

On April 15th, 2015, Nokia announced that it would acquire Alcatel-Lucent, a communication equipment vendor, by 15.6 billion Euros. The acquisition will reinforce Nokia's position in the communication equipment market. As scheduled, this acquisition will be accomplished at the beginning of 2016. The new company will use Nokia as its brand. This action mainly targets to the IT and CT integration and serves as the foundation for the development of Internet of Things and the transformation to IT cloud. After merging, the new company will have more than 40,000 R&D staff. The new company will accelerate its expansion towards new industrial direction which includes 5G, IP and SDN, cloud computing and big data analysis. In November 2015, Ericsson and Cisco declared they would establish a strategic alliance. These communication leaders' merger and alliances are the inevitable under the pressure of transformation in the Telecommunication 4.0 era.

In the new Telecommunication 4.0 market, the first type of traditional leaders comes from CT. They must enhance the IT capability and constitute an ecosystem to maintain their competence. With the reform of the CT industrial value chain, IT companies also hope to seize this opportunity to strike into the CT market. Furthermore, as the role of software integrators' importance is constantly increasing, chances that the traditional IT leaders extend their advantageous position to the CT market will increase and turn into the second type of leaders, the software integrators. Finally, the merger of IT and CT will probably form a third type of leaders, who will reform the overall communication industrial value chain.

2   New Industrial Ecosystem from Telecommunication 4.0: From Closed Industry Value Chain to Open Ecosystem

2.1   Changes to the Industrial Ecosystem: Customer Base Expansion in IoE and Diversified and Customized User Requirements

In the Telecommunication 4.0 era, the external environment and customer needs have changed a lot. According to the communication Maslow's Hierarchy of Needs, Telecommunication 4.0 is in the transition from information ubiquitous phase to perception ubiquitous phase or even intelligence ubiquitous.

At this stage, people have a strong demand for information collection and consumption, as well as diversified and flexible communication means. They hope to extend their perception by using external equipment.

This will lead to an IoE era, where connections between human to human, human to things or even things and things are realized. Equipment for connections will evolve from telephones to more diversified terminals (such as wearable devices and all kinds of sensors). The industry scope will become broader, changing from the original single industry value chain to interlaced nets of industries.

Under this new ecosystem, cross-border integration and mutual penetration between different industries will become more prevalent and unpredictable, giving rise to numerous innovative products and services. Correspondingly, the scope of network service customers will expand from traditional terminal users to IOT devices, or various practitioners of upper-layer products and services. These practitioners can further integrate with each other to deliver more services to terminal users via the open network of Telecommunication 4.0.

In the IOE era, surging service choices will bring more and more captious customers who have more rapidly changing demands. Customers will have more frequent tries for different services and higher requirements for ease of use. Lifecycle of a single service is likely to be shorter. Any product or service, once it can't fit into customers' needs, will be phased out quickly, since a large number of products and services are available in the new industrial environment.

In Telecommunication 4.0 era, cross-border integration will become common, and the closed telecommunication value chain will be replaced by an open industry ecosystem.

2.2 Ecosystem Competition: Enterprise Competition Becomes Ecosystem Competition, Build Your Own Ecosystem to Defeat Others

The traditional CT industry value chain is actually an ecosystem around operators. However, in Telecommunication 4.0 era, the concept of "IT ecosystem" reminds CT companies the occurrence of a new industrial regulation. For example, there are two types of cellphone OS, Apple's IOS and Open Source's Android. While there are also two editions of OpenStack, the Cloud OS in IT industry, one is an open source edition developed by many companies in the Open Source community, the other one is an in-house edition which is developed by each company for commercial release. Competition between these companies has evolved into competition between each ecosystem.

Huawei established its OPEN NFV open lab in 2015 to build up its own ecosystem. Huawei is leading integration test with other IT and CT vendors' products in order to grow their integration capability, while to make sure its own products can be integrated seamlessly to other system. Meanwhile, Alcatel-Lucent is also established an intact ecosystem with a large amount of cloud management, virtualization platform IT companies (like Red Hat) and other telecom service applications. Their objective is also trying to conduct system integration or to be integrated by other system integrators.

Some IT companies, such as HP, are also actively seeking their OPEN NFV partners. According to the NFV reference architecture, HP has certain strength in cloud management, orchestration, virtual machines and SDN controllers. As a result, its ecosystem establishment mainly focuses on network function vendors. HP is also actively collaborating with other hardware or cloud OS companies to provide its platform services into their solutions, while keep its own system integration capability.

Many successful cases in IT ecosystem are based on Open Source. But for CT vendors in the new Telecommunication 4.0 ecosystem, the major problem is how to embrace the Open Source ecosystem. Generally speaking, system integrators will support both Open Source's solutions (such as OpenStack) and non-open source solutions (such as VMware). By supporting two different solutions, they can select most suitable solution to customers with different requirements.

Everyone should know how to collaborate with different industries to fight for a larger ecosyetem.

One of the most fundamental cross-border cooperation in Telecommunication 4.0 is the cooperation between CT and IT companies. Operators and traditional CT vendors would like to form a win-win cooperative industrial alliance with IT companies. As mentioned previously, Ericsson invested in IT startups to improve its core competence.

Another kind of upper-level cross-border cooperation is target to network capability exposure and service integration. This type of cooperation is mainly between service providers or between service providers and operators. By integrating upper-layer services and bridging different service chains, an original singular product can be evolved into a diversified product, which can meet customers' various requirements via a single service experience. For example, a collaboration use case between a website and a supermarket. On this website, users can order some food while he is watching a food video on demand, without reading any advertisement. And the supermarket will initiate offline food delivery service once the user click the order button.

It is inevitable that in the future, competition between companies will become the competition between ecosystems. Companies must learn to organize their own ecosystem to take an advantageous position.

2.3   New Transformation Brought by Telecommunication 4.0: Product Operation to Service Operation

Telecommunication 4.0 not only brings great change to the CT industry, but also profoundly impacts operators themselves. It expands the position of operators and make higher demands. It pushes operators towards a brand new future in the following four aspects:

Firstly, Telecommunication 4.0 has brought operators unprecedented opportunities and possibility to control and operate the network independently. Before the Telecommunication 4.0 era, operator was just an organizers of the network. Network maintenance was mainly depended on equipment providers. With the appearance of Telecommunication 4.0, operators now can organize, deploy and maintain the network with automated tools, and become real masters of the network.

Secondly, Telecommunication 4.0 has broken the traditional competition rules between operators. The network itself is no longer the differentiation factors between operators. Only those operators who can provide differentiated services will become the final winner. Telecommunication 4.0 not only brings great flexibility, but also extensively simplified network deployment. This has deprived traditional advantageous operators the competition barriers and narrowed the network gap between them. Competitions among operators are even severe than ever before, which forces operators to shift from passive to active service. They have to get closer to users and serve them to gain advantages in future competition.

Thirdly, Telecommunication 4.0 has repositioned operators, who have switched from product operators to service operators. Before the Telecommunication 4.0 era,

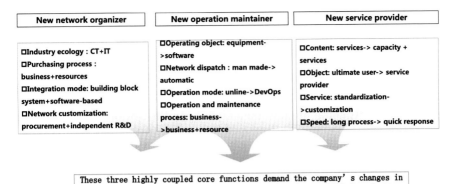

Fig. 4.6 Connotation change of operators' three role definition

operators provided users with simple network connection products such as simple languages, messages and flows. Now operators are no longer simple product operators, but service providers for people's lives and consumption. This requires operators to break off from traditional user-thinking and switch to customer-thinking. They should regard each user as a customer, actively focus on customers' requirements and provide new services.

Fourth, Telecommunication 4.0 has broken operators' traditional organization process. Telecommunication 4.0 has infiltrated into the production, establishment, operation, maintenance and service selling, accelerating the running speed of the whole process and caused the vague boundaries of the processes.

The traditional operators mainly played three roles, namely network organization builder, network operation maintainer and network product provider. In this new era, operators still hold the former two roles, but the connotation of these two positions is much more extensive than before. The third role has changed into the network service provider. With the constantly enriching of the three roles' connotation, the boundaries of the three grow vaguer and vaguer. We must abandon the traditional vision of the three role's definition, as shown in Fig. 4.6.

2.3.1 Operators, the New Network Organizer, Establish Network like Inserting the Lego

Before Telecommunication 4.0, operators are the simple assemblers of the communication networks. They organized the network by directly buying equipment from communication equipment providers and deploying corresponding network connection relations and then selling network products to users.

Telecommunication 4.0 has broken the traditional network equipment supply modes. By disassembling network equipment, it has introduced a new industrial ecological chain, including bringing in new virtual product providers, utility software providers, general hardware equipment providers, management software providers and powerful system integrators.

In the Telecommunication 4.0 era, operators are no longer simple network assemblers but builders of communication networks. They must assemble all the components into an integrated communication network, just like inserting the Lego bricks. In the network organizing process, operators must integrate the network with both software and hardware. After advanced planning and deployment of the hardware, organization of the software-based network will depend more on the design pattern and configuration script. Operators can easily match communication networks by designing patterns or flexibly compiling and dispatching the network flow direction with the network dispatcher.

In order to assemble and organize networks more effectively, operators need to guide the brand-new industry ecological chain and figure their role definition in it. They should find out cases where they should take the lead, industries with which they realize win-win situation and new network organization and building processes that are matched with the enterprise culture.

2.3.2   Operators, the New Network Makers, the Inner Gene of DevOps

Before Telecommunication 4.0, operators focused on the NE maintenance in the network operation and maintenance. The priority of the work is to monitor the network equipment, and the major method is simple network configuration. Be there a fault with the NE, operators usually order the equipment manufacturers to provide a remedy as soon as possible. When the network structure needs adjustment, operators usually change the network flow direction simply by some data configuration.

With Telecommunication 4.0, the network equipment can restore itself and NE maintenance is no longer the basis. The network equipment monitoring is also no longer the focus of the operation and maintenance work. The core has switched to the service-oriented quality assurance of the global business and the resource-oriented global resource control and optimal configuration.

In the Telecommunication 4.0 era, the past thinking that valued the stable operation and monitoring of the network should be turned into being network lifecycle management oriented. The DevOps processes will be more frequently embedded in the operations. It will be particularly important to use the operational tools proficiently and have real-time control over the network. Meanwhile, with the decoupling of the software and hardware, the business and telecom cloud IT infrastructure oriented operation and maintenance are relatively independent and also need full cooperation so as to obtain the management synergy capability. This has proposed higher requirements for operators, who must possess the new maintenance capabilities, the DevOps strength and the double-process and double-management abilities.

In the Telecommunication 4.0 era, not only operators' operation methods have changed, but also the network operation state. A new normal has been introduced. It is mainly reflected in three aspects. Firstly, it is an association of activity and inertia. In this era, the communication network has changed into a software-based network. On condition that the network-wide computing, storage, and network resources remain relatively stable, the entire network topology and network capacity can be dynamically adjusted in the form of computer rooms, regions, provinces or even the whole country. The work focus of the operators has transformed from the previous focus on static network planning to the combination of static resource planning and dynamic network adjustment. Thus, the association of activity and inertia has become the new normal for Telecommunication 4.0. Secondly, it is globally and partially interwoven. By supporting the dynamic network adjustment and network capacity auto stretch, it has become possible for the global scheduling and real-time optimization of the network. At the same time, operators can take the overall situation, start locally and link comprehensively. This global and partial intertexture has become the new normal. Thirdly, it is both rapid and stable. In this era, the software-based network implementation model has enabled the rapid upgrading of the network deployment. The networks respond to the market demand quickly. It is even possible to update to a new version every day. But at the same time, future network development calls for a stable network core competence and a rapid new service supply. Telecommunication 4.0 can support these two demands and realize the combination of rapid service supply and stable network competence. The above three new normal are also interlaced and mutually influencing, jointly promoting the development of the operators' networks.

2.3.3 Operators, the New Network Service Sellers, the Endogenous Capability of NaaS Becomes

Before the Telecommunication 4.0 era, operators were not equipped with the endogenous opening capabilities, only limited opening capabilities. Traditional communication openness relied on independent platforms, which served between the network and applications. There were mainly two modes. The first is to acquire capacity from the network with the communication agreement, opening to the outside applications with the application-layer interface protocol. The other is integrating network capacity and opening with the Internet open protocol.

With Telecommunication 4.0, the change of network mechanisms has also varied the service supply modes. Operators can quickly integrate the network capacity, deeply customize the service features and flexibly offer services to users. Their services are no longer restricted to the traditional ones. The network capability has also become a special service to offer the third-party developers, making NaaS possible. Besides, Telecommunication 4.0 has turned openness into an endogenous capability of the operators. The specific opening platforms are no longer needed, as reflected in the following three aspects in Fig. 4.7.

Fig. 4.7 Endogenous open
mode

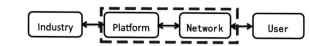

1. **Programmable network direct openness**

By opening the interface technology of network controllers, users of the Apps can conveniently select the network bearer channel and directly use the network resources, making the direct openness of the net possible.

2. **Arrangeable function direct openness**

With Orchestrator, Telecommunication 4.0 can provide an open interface for the upper-level Apps, which can in turn create new services with the help of application arrangement and scheduling network functions. The network open interface can guarantee the quality of end-to-end services and quickly turn upper-level App providers into MVNOs.

3. **Directly opening infrastructure**

The infrastructure platform of network function in Telecommunication 4.0 can be built on the telecom integrated cloud platform. This kind of cloud infrastructure can serve as the third-party application for deploying infrastructure which can be obtained by users through communication networks.

2.3.4 The Three New Roles Obscured Boundaries and Formed a Rapid Closed-Loop

Before Telecommunication 4.0, the three roles had clear division interfaces and each performed its own functions. When there were new business requirements, it usually took years from requirement proposal to network regulation and organization, to network operation and maintenance and to bring the service online and supply users.

Now with the introduction of the DevOps mechanism, operators' roles in network service supply, network organization, operation and maintenance grow vaguer and vaguer. The three roles have formed an intertwined close relation and in turn evolved into an even closer mutually promoting and coordinated rapid closed-loop in the work flow. This has enormously shortened the launching time for a new service, simplified the network organization and facilitated network operation and maintenance.

In order to adapt to the new role definition, the mainstream operators all over the world have made all kinds of attempts in the enterprise transformations. These operators can be divided into two types in summary. The first type chooses to actively hug the Telecommunication 4.0, initiatively explore enterprise transformation and seek for industrial technological change, such as the AT&T, Telefónica of Spain, DoCoMo of Japan and CMCC, etc. AT&T has put forth the Domain 2.0 strategy. By 2020, virtualization controlled by software will have covered 75% of the network. The Open source software proportion will increase from today's less than 10–50% in 2020. It is also endeavouring to transform into a software enterprise by 2020. CMCC released a network vision named NovoNet in July 2014. It has technically put forward its consideration and requirements for the future network. In order to realize the NovoNet vision, CMCC is actively exploring the operator transformations in terms of enterprise culture and organization structure. Technically, it has launched the NFV pilot, deploying SDN technical solutions in DC and restructuring operators' communication rooms through the telecom cloud.

The second type of enterprises chooses to follow closely. They actively try new technologies, but lack the initiative to explore mechanical transformation and institutional reform, such as the Orange of France, Vodafone, Texas Instruments, etc. These enterprises usually trace and participate in some technology reform by expanding the current R&D team. At the same time, they also actively carry out laboratory tests and vigorously advertise and promote their strategy in this direction.

No matter which kind of transformation, we believe that all operators need to make deep-level enterprise transformations on their way to Telecommunication 4.0. NFV/SDN is a new means of production. As the producer, operators must level up and develop new productive relations to use the new production tools and improve the productivity effectively.

Chapter 5
Telecommunication 4.0 Opens New Mode, New Space and New Development of the Future

Telecommunication 4.0 represents a new development phase of communication, and at the same time the information and communications industry as well as other relevant industries will undergo profound changes.

Firstly, mobile communication will enter the 5G stage, where the whole world would be closely connected, and information would be exchanged easily and widely.

Secondly, the internet is entering the "Internet Plus" era, and the combination of the internet with all walks of life will usher in tremendous space for development.

Thirdly, the transformation and upgrade of manufacturing will focus on quality rather than size under the macro-planning of "China Manufacturing 2025".

This great plan is giving the birth to a new communication network, and this communication network should support and serve the development of new strategies. It can be predicted that Telecommunication 4.0 and 5G Technology will complement each other to seek common development, go hand in hand with Internet Plus to explore a promising future and play key roles in "China Manufacturing 2025" to support the upgrade and transformation of the manufacturing in our country, described in Fig. 1.

1 Telecommunication 4.0 & 5G: 1 Plus 1 Is More Than 2

1.1 Overall Vision of 5G

While 4G has enabled people to enjoy high-speed internet services and has profoundly changed their lives, 5G is in urgent need of the explosive amount of mobile data traffic, numerous device connections and emerging businesses and application scenarios.

© Springer Nature Singapore Pte Ltd. 2018
Z. Li, *Telecommunication 4.0*, DOI 10.1007/978-981-10-6301-5_5

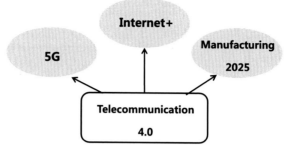

Fig. 1 Overall Relationship

5G will penetrate into all fields of life, and a user-centric information ecological system will be established. 5G technology, enabling information to break through the obstacles of time and space, will provide users with a wonderful interactive experience and feast of personal information; 5G technology will narrow the distance among people, achieve smart connectivity between human beings and the whole world through seamless fusion, shown in Fig. 2.

5G will offer an optical fiber-like access rate and zero latency user experience, including the consistent services in the scenarios of the ability to connect hundreds of millions of devices simultaneously, ultra-high traffic density, connection density and mobility. The intelligent optimization of services and user awareness would enhance the network with the increase of energy efficiency by a hundred times, reduce the bit cost by a hundred times, finally achieve the overall vision of everything connected closely and information exchanged easily.

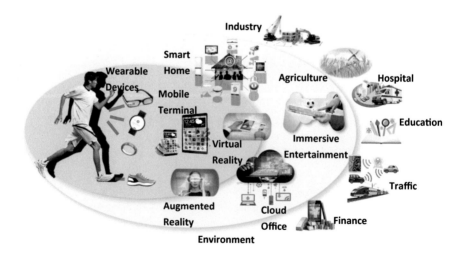

Fig. 2 5G vision

Generally speaking, 5G will penetrate into all walks of life. Based on mobile broadband, 5G will help to establish a flexible and dynamic network to meet the requirements of different industry.

1.2 Main Engine of 5G Development

Mobile Internet and the Internet of Things are two main growth engines of mobile communication development in the future. 5G will have a promising future.

Mobile internet has subverted the traditional mode of mobile communication services, which offers users an unprecedented experience and profoundly changes people's lives. Looking into 2020 and the near future, the popularity of UHD, 3D and immersive videos will substantially drive up the data speed. For instance, after the 3D-8 KB video is compressed into a hundred times, its transmission rate is 1 Gbps. Augmented reality, cloud desktop, online games and other services not only pose a new challenge to the transmission rate of upstream and downstream data, but also require "zero perception" of the delay.

In the future, there will be large amounts of personal and office data stored up in the cloud, the interaction rate of large amounts of real data can be comparable with the transmission rate of optical fibers, and it will bring the pressure to the mobile communication network in the hotspot region. The OTT services like social networks will be one of the dominant applications in the future, and the frequent use of small data packages will cause massive consumptions of signaling resources. In the future, people will require more and more for the communication experience under many different application scenarios—they expect high-quality services in some populous venues like gyms, outdoor gathering and concerts, as well as high-speed environments like railways, automobiles and metros. Further development of mobile internet will drive the tremendous increase of mobile traffic by thousands of times, thus promoting the technical and industrial reform of mobile communications.

The Internet of Things has expanded the scope of mobile communications' services, from people-to-people communication to thing-to-thing and people-to-thing smart connectivity, making mobile communication technology available in more industries and fields. In the coming future, mobile health care, IOV, smart home, industrial control and environment monitoring will promote explosive growth of applications of the Internet of Things, and hundreds of millions devices will be connected to the internet, thus "connectivity to everything" will finally be achieved, and large-scale new industries will be born to vitalize mobile communication.

The Internet of Things (IoT) has numerous types of services, which are fairly different. The services like smart home, smart grids, environment monitoring, smart agriculture and smart meter reading require the internet to support large amounts of device connections and frequent use of small data packages; video monitoring and mobile health care demand higher transmission rates; services like IOV and industrial control require millisecond level delays and close-to-100% reliability. Moreover, lots of IoT devices are located in remote areas like mountains, forests and near bodies of water, as well as corners in the house, basements and tunnels where signal is hard to reach, so these require the mobile communication network to expand its coverage. In order to cover more services of the Internet of Things, 5G technology should be more flexible and scalable to adapt to abundant device connections and various needs of the users.

In addition, the limitation of extending 4G urges 5G to develop rapidly. In the 4G network, frequent introduction of new functions forces the network to put patches, which makes it more and more redundant, being fallen behind the rapid development. Therefore, people expect 5G to have a brand new network architecture, efficient network management, lower operation costs and open network capability. The 5G network should be redesigned from its root to create an optimized network system, and integrate with the internet effectively.

The 5G services demands, network demands and new technologies represented by NFV/SDN jointly drive the innovation of 5G network architecture to support diverse access scenarios, to satisfy the end-to-end service experience, to achieve flexible network deployment and efficient network operation.

1.3   Telecommunication 4.0 and 5G Complement Each Other to Seek Common Development

As shown in Fig. 3, as a part of Telecommunication 4.0, 5G and Telecommunication 4.0 complement each other. On the one hand, the 5G network will cover all fields of society in the future. Based on mobile broadband, a flexible and dynamic network

Fig. 3   Telecommunication 4.0 and 5G complement each other

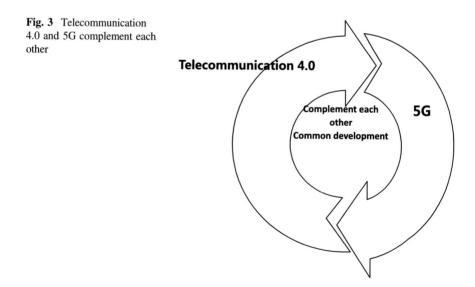

will be established to satisfy the demands of different sectors. And the features of Telecommunication 4.0 could meet the needs of the core development of the 5G network. Therefore, Telecommunication 4.0 is the core fundamental of the 5G network. In the future, the 5G network may utilize the network resource of Telecommunication 4.0 to build a flexible and efficient mobile communication network. On the other hand, Telecommunication 4.0 is not confined to 5G network, it can be developed further based on the 4G mobile network and fixed network.

Telecommunication 4.0 and 5G should achieve the common development. Through the redesign of its functions, the 5G network could obtain stronger network capability than the former generations of mobile communication, which can achieve customized services, modularized functions, virtualized devices and unified management. Moreover, it enables the mobile network to change from rigidity to softness, from CT to ICT, from single network to the coordination of integrated smart network and distributed management network.

2 Telecommunication 4.0 and "Internet Plus": Towards an Unprecedented Future

The *Opinions on Actively Promoting "Internet Plus" Action* enacted by the State Council is influential to all economic and social fields in our country in 2015. The goal of "Internet Plus", based on the rapid development of the internet, is to further integrate the internet with all economic social fields, to promote the formation of emerging businesses, to drive economic growth, to encourage mass entrepreneurship and innovation, and to achieve the joint development of network economy and real economy. The development of "Internet Plus" raises a fresh issue on commercial activities of the communication industry. To develop and innovate Telecommunication 4.0, the main task is to develop communication technology, system and structure, innovation mechanism, development and operation mode as well as the engineering system that can support the "Internet Plus" development mode.

2.1 "Internet Plus" Explores the Path with Chinese Characteristics in the Third Industrial Revolution

In March 2015, premier Li Keqiang proposed in the *Report on the Work of the Government* that China would make an "Internet Plus" Action Plan. In July 2015, *Opinions on Actively Promoting "Internet Plus" Action* was officially announced. According to the development goals set by the State Council through the "Internet Plus" action plans, internet and social economic development would be further integrated. Internet-based new business would become the new engine of economic development. The internet would play an important role in supporting mass entrepreneurship and innovation, becomes the major means to provide public services, and the joint development of network economy and real economy would emerge. By 2025, the new economic form of "Internet Plus" would be formed, and "Internet Plus" would become the main driving force of social economic innovation and development in our country.

From historical perspectives, we find that all industrial revolutions, including the first one represented by the power innovation of steam engines to replace manual tools by machines, the second one represented by power technology innovation to achieve the wide use of energy, and the third one in the 20th century represented by the innovation of atomic energy, computer technology and space technology to achieve the wide use of information control technology, have promoted social economic innovation and development by increasing new elements of market resources allocation, transforming the means and structure of production and resource allocation, or in other words, by changing the production mode and productive relationship.

Based on this judgment, the basis of the theory of "Internet Plus" can trace to a book written by Jeremy Rifkin: *"The Third Industrial Revolution: How Lateral Power Is Changing the World"*.

Rifkin reflected deeply on social economic changes brought by three industrial revolutions. He found out that many economic revolutions in history just occurred right in the combination of new communication technology and new energy systems. Moreover, this kind of combination will come again—namely, the combination of internet technology and renewable energy. And it builds a new and strong infrastructure for the third industrial revolution. In the 21st century, the third industrial revolution will fundamentally change people's lives and work.

The view of treating the internet as the driving force of social economic innovation and development is a big theoretical leap of the third industrial revolution. The view promoting social economic innovation and development through the internet comes from two important practices of the information and communication field.

Mobile internet is the first key practice of the information and communication field. The establishment of a mobile internet ecological environment represented by iPhone and Android Phone is its milestone. Now, mobile internet is able to meet various demands of the public, support mass innovation, and encourage the communication network, mobile device, operating system and application development to form a sustainable ecological environment. According to the statistics from CTIA, the intelligent terminal coverage in America extended 50% from 2009 to 2013. By advocating an innovative ecological environment, the development cost of applications has been significantly reduced. The development of mobile internet applications has formed a new industry worth over a million dollars, which has created plenty of jobs.

The development of mobile internet directly drives the telecommunication network to progress from 2.0 to 3.0, and the commercial activity of telecommunication operations evaluates from stock operation to increment-based traffic operation. In

2013, American operators invested $97 per household in the 4G LTE network, about four times than the world average, and this figure grew to a record $170 in 2014. At the same time, the application of mobile internet drives the development of intelligent hardware like intelligent terminals and wearable devices. It also drives rapid progress of cloud computing technology, big data technology and artificial intelligence technology, and achieves the implementation of voice and image's signal processing.

Mobile internet appeared in China in 2010. China Mobile, together with HUAWEI and other communication device manufacturers and terminal development manufacturers, develops and deploys the fourth-generation broadband mobile communication network and intelligent terminal of TD-LTE. Internet information and consumption service providers like Baidu, Alibaba and Tencent are all on the rise. Through all these above, mobile internet has profoundly changed people's way of life and communications, and commercial activities of traditional communication, media and retailing industry have completely changed.

The second important practice in information communication is the internet-based informatization of enterprises, governments as well as public departments. The enterprise is the main body of the market, and market competition is the main impetus of enterprises' informatization. However, over the long term, due to the limitation of technology, management modes, safety requirements, market patterns and some other reasons, enterprise IT (including the enterprise internet portal, ERP, CRM, MES, OA, etc.) has been in the independent development system. And the information system for the departments of trade management and public services has also been independent from the internet for a long time. Nevertheless, thanks to the development of cloud computing, security technology and mobile internet, this pattern has been transformed significantly in recent years.

On the one hand, the information system of enterprises and governments is transferred to the cloud computing platform through visualization, their portals, customer service and customer relationship management change from the application software in the enterprise server to cloud service established upon mobile internet. On the other hand, mobile internet terminals achieve cloud connection, personal terminal equipment of users can be used in BYOD (mobile officing), and the abundant public data controlled by governments begin to be open to the public and enterprises. The practice in this field lays a solid foundation to integrate the internet with enterprises, governments and public departments, so as to promote its further integration with all social economic fields.

Guided by the theory of the third industrial revolution, and based on the development of global mobile internet in the information consumption field as well as the practice of integrating the internet with the information system in enterprises, governments and public departments, we should further integrate the internet with all social economic fields, treat new businesses of the internet as a new engine of the economy, support mass entrepreneurship and employ the internet as an important means in public services, so as to achieve the interaction between network economy and real economy. And this is a grand innovation in the economy with Chinese characteristics that the government has proposed in the development wave of the internet. In this wave of innovation, the information and communication industry is playing a key role. It requires new techniques, network mechanisms and business modes to be upgraded from Communication 3.0 to Telecommunication 4.0, then to help form the "Internet Plus" economy and make "Internet Plus" a main driving force of social economic innovation and development.

2.2 Deep Integration with Social Economic Fields Is the Foundation of the "Internet Plus" New Businesses

According to the "Internet Plus" Action Plan released by the State Council, the most important symbol is to integrate the internet with all social economic fields and employ the internet as a main driving force of economic development. This also means that "Internet Plus" and information consumption of mobile internet faced consumers [B2C (Business to Consumer) or C2C (Consumer to Consumer)] are essentially different.

Deep integration produces new businesses, which can be an incentive to economic development, covering manufacturing, agriculture, national defense, science and technology, education and people's life. Apart from that, internet infrastructure and its operation mode which support promising "new businesses" of the information and communication industry will also undergo profound changes.

The "Internet Plus" Action Plan released by the State Council tends to promote transformation, integrate innovation and public concerns. Therefore, the first eleven-specific-action has been published.

Entrepreneurship and innovation based "Internet Plus". Internet should provide strongly support to entrepreneurship and innovation to promote gathering, opening and sharing of all kinds of elements and resources and cultivate a favorable environment for mass entrepreneurship and innovation.

Collaborative manufacturing based "Internet Plus". Develop intelligent manufacturing and large-scale personalized customization to enhance the networked collaborative manufacturing and accelerate the servitization of manufacturing.

The modern agriculture based "Internet Plus". Establish the internet-based agricultural production and operation system, develop precise production modes, and cultivate diversified and web-based service modes.

Smart energy based "Internet Plus". Promote energy production and consumption intelligently, establish a distributed energy network, and develop grid-based communication infrastructure and new types of businesses.

Inclusive finance based "Internet Plus". We need to promote the establishment of internet-based financial cloud services platforms, to encourage financial institutions to broaden service coverage through the internet, and to expand the depth and width of internet financial services.

People-oriented services based "Internet Plus". Innovate web-based management and services of governments, develop online and offline new consumptions and internet-based new services, like medical treatment, health care, old-age care, education, tourism, and social security.

Efficient logistics based "Internet Plus". Establish information sharing and connecting system of logistics, build up intelligent warehousing systems, and improve intelligent logistics distribution and deployment systems.

E-commerce based "Internet Plus". Develop e-business in rural area, industry and cross-border, and promote innovation of e-commerce applications.

Easy traffic based "Internet Plus". Enhance the internet level of traffic infrastructure, transportation and operation information, and innovate easy traffic services.

The green ecology based "Internet Plus". Promote the deep integration of internet with ecological civilization, intensify dynamic monitoring of resources and the environment, and achieve connectivity and sharing of ecological environment data.

Artificial intelligence based "Internet Plus". Accelerate the core technology breakthrough of artificial intelligence, cultivate and develop new industries of artificial intelligence, promote innovation of intelligent products, and enhance the intelligence level of terminal products.

Obviously, the eleven specific actions of the "Internet Plus" Action Plan proposed by the State Council has built up a basic frame and access route of new businesses to integrate the internet with all social economic fields. It also points out a clear demand framework and development path of information and communication technology, system structure and business mode of innovative new businesses which support "Internet Plus" development.

2.3 Grasp the Features of "Internet Plus" New Businesses, and Promote the Development of the Telecommunication 4.0 System

We have to be aware that the new businesses produced by "Internet Plus" include the information and communication industry, which is quite different from the traditional industries. There are several features as below.

Cross-domain businesses based on context may simultaneously integrate with some independent industries. Automobile manufacturing, automobile products, roads, traffic management and traffic transportation were separated industries in the past. However, based on the context relationship among "internet + automobile + transportation", the new business of car–road networking includes product modality of automobile manufacturing (intelligent networking cars), production mode of automobile manufacturing (software-defined cars), usage mode of automobiles (transportation mode), maintenance services mode of automobiles (full life-cycle services), transportation control and management modes (Intelligent Vehicle Infrastructure Cooperative Systems), as well as the operation modes of transportation and logistics. It also includes the innovation of technology, architecture and business model of automobiles, traffic and transportation.

The new businesses transform the supply chain centered on products or services from arborescence to planar netted interrelationship, in order to forming a client-centered full life-cycle industrial ecology. New businesses have turned an enterprise-based supply-demand trade into a multi-stakeholders-based diverse and coordinated ecological environment. For example, through "Internet Plus", the arborescence supply chain of automobile products including mechanical parts, electronic components and software can be turned into a coordinated ecological environment which has netted interrelationship. On the one hand, the automobile can be implemented as software-defined to enhance the traffic safety, transportation efficiency and reduce the environmental pollution. On the other hand, it can promote continual production and innovation of overall products and components of automobiles.

"Internet Plus" new businesses are established upon the fast-growing information and consumption internet, like internet information services, internet media, social media and internet e-commerce. They gradually expand based on the information and consumption internet. On the one hand, the expansion needs to consider the maturity, complexity and innovation space of its integration with related fields. On the other hand, it is also related with the maturity of information and communication technologies that support the integration, as well as the stimulation for the innovation and development of the information and communication industry.

New businesses produced by "Internet Plus" integrate not only with social economic fields, but also with industrial supervision and public services of the government. In the public services sector, like public security, the integration of industrial supervision, public services and other related economic fields raises many innovation requirements and solutions on the market level. A typical example is deep IOV network. It is a new market aimed at ensuring people's traffic safety and avoids 80% of traffic accidents by controlling automobile motions. Due to this technique, the former market of automobile based on vehicle condition diagnosis, active safety and assist driving systems progressed into a new market based upon active automobile safety and road transportation safety, which will further move forward to autonomous driving; it dramatically reduced the traffic regulatory cost of the government, and the level of traffic public service has been largely enhanced. What's more, it helps traffic-related industries, like insurance and business traffic, to join the market and benefit from it.

Obviously, with the development of "Internet Plus", new businesses of the information and communication industry which support the promising and widely-integrated "Internet Plus" new businesses will confront with an unprecedented pattern where more and more "Internet Plus" applications will pose new challenges to the infrastructure and operation of the internet. This also means that the traditional network system which is based on a standardized network and user interface and aimed at providing information and consumption services, together with cascade telecommunication operation modes consisted of technology development, products development, networking and products will be thoroughly overturned by new technology systems and operation modes. Therefore, developing a "Internet Plus" engineering system is in urgent need.

Telecommunication 4.0 is born to meet the needs of the innovation developing mode, "Internet Plus", to communications. To support the deep integration of the internet with all economic social fields, Telecommunication 4.0 meets the increasing demands of capacity and bandwidth. It also needs to promote new technology, establish new architecture, form new business modes and engineering systems. Telecommunication 4.0 proposed network function virtualization, software-defined networks and intelligent hardwares are the technical starting point of innovative development modes. The open system, software-defined network functions, resource-sharing and programmability of the network are the basic requirements. Establishing new development systems and business modes should learn from the experience of mobile internet ecological development, especially "using bandwidth, computing and storing capability to exchange the complexity of integration and innovation", an initiative gene of the internet. The experience also includes building innovative development modes for Telecommunication 4.0 to support "Internet Plus" based on the "iteration" and "join" modes of mobile internet, and establishing an innovative and multi-beneficial ecological environment based on self-adaption, self-organization and prompt plug and play.

3   Seize Opportunities, Face Challenges, Transform Actively to Promote the Development of Telecommunication 4.0

As a new-generation network, Telecommunication 4.0 is a tremendous revolution for the communication industry. The big change leads to the upgrade and update of industries and heavily emancipates the productive forces. Now, our country is on the critical stage of new normal and industrial transformation, so Telecommunication 4.0 has the epoch-making significance.

Telecommunication 4.0 is the key to the upgrade and update, industrial transformation as well as the development of changing from extensive to intensive. For now, China has entered the new normal, where high-speed economic growth is replaced by medium-high speed. Economic structure is optimized and upgraded, and investment-driven economy is turning into the innovation-driven economy. After ten years' growth, Chinese communication industry has also entered the new normal, where extensive industrial growth relied on huge investment turned into intensive growth relied on total factor productivity. However, there are some internal problems troubled Chinese telecommunication industry, like being relatively closed and lacking ability in critical links. The transformation of

Telecommunication 4.0 could be utilized to solve them on the new platform. Innovation has always been the motive power of the development of the communication industry. Therefore, we should drive the innovation and actively carry out the reform.

Telecommunication 4.0 is the key for Chinese communication industry to reinvent itself and serve information society better. At present, some national strategies, like "Internet Plus", "China Manufacturing 2025", "Mass innovation and entrepreneurship", have more requirements on the communication industry. The communication network, as an important information infrastructure of China, should better fit in, support and serve the national strategy. This is the requirement of its development, and also a good chance to remodel itself.

Telecommunication 4.0 is an important moment for Chinese communication industry to follow the information era and redefine the global competition pattern. Now, in the Information 2.0 era, both globalization and flattening are accelerating, and global competition pattern becomes more complex. On the one hand, Chinese communication industry should forge ahead with determination, employ new technology to enhance the competitiveness and soft power, take advantage of the ICT-integrated new technology to create new platform and improve the global competitive-strength. On the other hand, our communication industry should make full use of favorable strategies, like "the Belt and Road" and infrastructure construction, to rebuild the global layout. Utilize the infrastructure of new communication to join global competition from a high starting point.

Three generations of communication, from 1.0 to 3.0, each experience is an arduous and self-broken process. Telecommunication 4.0 is a long road to take, there are several issues should be focused on as below.

1. **Focus on the industrial reconstitution of Telecommunication 4.0, overturn and being overturned coexist**

In the Telecommunication 4.0 era, IT and CT are integrated, which replaces the old industrial order. Therefore, building the new industrial order is necessary, and some traditional industries will be replaced by the new industries. The coexistence of overturn and being overturned poses great challenge to operators and manufacturers.

Challenges still exist: How do operators use the concept of Telecommunication 4.0 to transform from the traditional one to the new one? How do manufacturers seize the chance of Telecommunication 4.0 to upgrade products and explore new business opportunities?

For operators, their lack of software genes makes it difficult to introduce IT. Operators have challenges to import IT in several aspects, like talents, culture, mechanism, etc. Especially, Telecommunication 4.0 is the best time for operators to reintegrate the industrial chain and obtain stronger competitiveness, and otherwise they will descend to be pipeline operators.

For manufacturers, they confront with heavier challenge and impact. Before Telecommunication 4.0, the industrial chain of communication manufacturing was single, and the core players were equipment manufacturers and chip manufacturers. After entering the Telecommunication 4.0 era, however, the core players of the new industrial chain will include general hardware manufacturers, chip manufacturers, virtual software suppliers, NE functional software suppliers, management equipment suppliers, etc. The traditional equipment manufacturers will suffer more impact. The former integrated selling mode of hardware and software was decoupled into three layers: general hardware, virtual software and NE functional software. Traditional equipment manufacturers have strong technical barriers in NE functional software, and they also face fierce competition from the IT field in general hardware and virtual software.

Traditional equipment manufacturers of communication and newly-introduced IT equipment manufactures will have head-on competition in the Telecommunication 4.0 era. With years of development, there are some world-level enterprises, including HUAWEI and ZTE. In Telecommunication 4.0 era, we should pay high attention to the anticipated reshuffle and new competition pattern.

2. **Establish the core capability in Telecommunication 4.0 era, especially the software ability**

One of the essences of Telecommunication 4.0 is software-based communication network, including the realization of software-based network, and software defined network (including orchestration) of the network. The abilities centered on software ability, integration ability, and open source ability become core abilities of Telecommunication 4.0.

From the view of "manufactured in China", the communication industry is presenting a trend of "the harder, the stronger; the softer, the weaker". "The harder, the stronger" is shown by the good system design, strong integration ability. The hardware system integration and board card design of China-made equipment almost can be done by self-research. "The softer, the weaker" is shown in software. Although management, security, accounting, OA and other applications have realized the self-developed, the design and control-ability of basic software like virtualization platform, operating system and database are very weak. Now, the operating systems and large-scale relational database products which cannot be self-developed mainly rely on imports.

Telecommunication 4.0 proposes requirements on the self-control of traditional operating system and virtual platform as well as new-type network operating system platform, like controller platform and orchestrator platform. We have to deepen the reform and change the backward situation of software with the help of Telecommunication 4.0.

Compared with the hardware system, software system is concerned with more open and complex ecological chains. And the open source mode is the foundation of IT software operation. The communication industry should learn from IT software development mode to extensively rely on the open source. Open source is not only the opening of code, but also the opening attitude and culture.

Traditionally, open source is referred to uploading the realized code to open source community. However, this is not enough. How to attract more developers to join the project, and how to build a community to develop the open source code are the culture that we need to deeply understand. In this point, management staff and

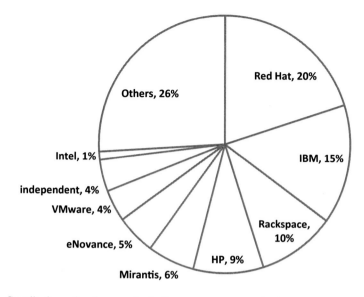

Fig. 4 Contribution ratio of companies in the open source organization

developers of Chinese enterprises need to understand the open source culture, respect developers, cultivate capable open source engineers and build dynamic open source communities.

Honestly speaking, we do not have a solid foundation of open source, even not understand its culture. Our contributions to the international open source communities are limited. The most popular foundation—Linux has not seen any independent project led by Chinese companies, only have a few joint projects. In the most successful organization—OpenStack, only a few Chinese companies have made contributions, shown in Fig. 4.

In the Telecommunication 4.0 era, with the introduction of open source software, open source should be supported in the national strategy level. In addition, large-scale enterprises, universities and research institutions should be driven to contribute more in open source to enhance the soft power of China.

3. Focus on the cultural change of Telecommunication 4.0, the key is to break the old ways of thinking and establish the new

Telecommunication 4.0 is an era of self-control, which is more open, tolerant, global, and diversified. Industrial integration, technological improvement and industrial upgrade will certainly lead to huge development and transformations.

To fully make use of Telecommunication 4.0, we should get rid of the experientialism, bringism and dependencism constrained the development. Employ open-source thoughts and industrial operation modes of IT to establish new software thoughts, overall-situation thoughts, DevOps thoughts, global thoughts and awareness of self-innovation. We should also promote industrial cooperation with the open mind.

(1) Build up the overall-situation thoughts to promote Telecommunication 4.0

Businesses of the traditional communication industry were not closely connected, which led to the interior barriers of the industry and impeded industrial development and network optimization. Telecommunication 4.0 has achieved a public platform and unified orchestration to facilitate overall resource optimization, which forms a network of complete open, overall optimization and breaking the traditional geographic restrictions. Establishing the overall-situation thoughts is to properly handle the relationship between "all" and "part", "reality" and "development", and to employ the overall, integral and visional thinking mode.

(2) Build up the DevOps thoughts to promote Telecommunication 4.0

Telecommunication 4.0 is a dynamically adjusted network whose functions are continuously overlapped. And it requires DevOps to accelerate the supply of businesses of operators and the deployment of the network. However, when we look at the global network operators, the existing staff skill and organizational structure cannot support DevOps. Transforming to DevOps is concerned with the mechanism, while the cultural gene and people are more important: the transformation requires the retraining for the employee, the introduction of new-type employees and the remodeling of corporate culture.

(3) Build up the global thoughts to promote Telecommunication 4.0

Like the development of economic globalization, the whole world has been narrowed into a small global village. The era has gone that a single company or a single country completed the whole industrial chain of R&D, production and sales. Instead, a complete chain requires the cooperation from many companies and countries. We should make full use of all advanced technologies and industries. Traditional communication equipment manufacturers should abandon the former development mode. By cooperating with foreign manufacturers on general hardware platforms and virtual software can enhance the competitiveness of their own products. Domestic communication operators should make full use of the global industrial chain to build an industrial ecological environment.

The overall-situation thoughts, DevOps thoughts, and global thoughts cannot be formed by a simple slogan, and the change of thoughts needs the remodeling of employees' ability and quality as well as the support of corporate culture. These are the decisive factors of competitiveness of the enterprises.

(4) To layout the Telecommunication 4.0 era in advance, and choose our own path

The development of Telecommunication 4.0 should obey the feasibility principle to adapt the national conditions. For example, beautiful crystal shoes can look gorgeous only on the right feet.

Operators can introduce Telecommunication 4.0 in two ways: overturning introduction and gradual introduction. The core concept of overturning introduction is to establish the new Telecommunication 4.0 network and gradually replace the existing one. However, the gradual introduction means that carrying out transformation by step to support new technologies like NFV and SDN based on the existing network equipment. The demand of Overtruning introduction is much high to operators. The overall planning and detailed implementation steps should be completed in the initial stage. And the risk of the integral moving of businesses during implementation should also be considered. It is pricy and risky in the initial stage, while the overturning introduction could rebuild the operator's network and solve problems of unreasonable legacy network planning and difficulties of optimization. The gradual introduction, which has low requirements, is essentially a mode of "Crossing the river by feeling for stones", where businesses are gradually transformed and finally integrated into the unified Telecommunication 4.0 network. It is a process of trial and error, while it cannot use Telecommunication 4.0 to resolve an unreasonable legacy network plan.

Choosing the proper development path of Telecommunication 4.0 based on their own conditions is the right choice. In other words, instead of copying the experience of other countries, enterprises should consider their own conditions before changing corporate culture, rebuilding the organizational structure and training the employees. Only by learning, digesting and obtaining can truly develop.

New productive relations should be in line with the new productivity—Telecommunication 4.0 to fully use their parts. Therefore, the entire process needs to be optimized to adapt the technological development, and the organizational structures of companies need to be changed as well. However, transformation of the productive relations cannot be achieved overnight, and it must be restrained or even obstructed by the old one. How to adjust the organizational structure and optimize the entire process needs their wisdom, their expectation and determination for the future.

Overall, the Telecommunication 4.0 era is a new and great time for the communication industry. The new era raises new development requirements, which offers chances for the development of the communication industry. Telecommunication 4.0 has brought new opportunities to operators, equipment manufacturers as well as the upstream and downstream of industrial chains; it is also a good time for the Chinese communication industry to achieve "curve overtaking", promote software ability, and change the industrial development mode. Now the communication industry is in the stage of germination, and we are on the same starting line with foreign countries. Moreover, our voice is better heard by the international community. We should make full use of this window stage of development to overcome difficulties, promote innovation, transform actively, and support our country to progress from a big communication country to a strong communication country.

Chapter 6
Telecommunication 4.0: Opportunities and Challenges

1 Finding New Dividend in the Era of Telecommunication 4.0

Zhilei Zou, MNO BG President, Huawei Technologies

From analog telecommunication over 100 years ago and digital telecommunication in the 1980s to the Internet, which saw explosive growth in the early 21st century, global telecommunication has gone through 3 eras—telecommunication 1.0, telecommunication 2.0 and telecommunication 3.0. In these eras, consumers are provided with cheaper and better telecommunication services and the global telecommunication industry is becoming increasingly prosperous.

When it comes to the second decade of the 21st century "it is the best of times, it is the worst of times". This best describes the telecommunication industry now. On the one hand, the Internet based on traditional telecommunication is having a profound impact on the world and human beings. Our senses are infinitely extended and expanded through the Internet. People have easy access to the digital world, significantly changing their life style and business models. A new digital era is coming to us. On the other hand, after 30 years of prosperity, the telecommunication industry faces obstacles to further develop. The expanding telecommunication networks now cover most people in most markets, making the traditional mode in which development mainly relies on the growth of users unsustainable. Meanwhile, with the development of mobile Internet,operators' profit from network business are badly affected by OTT enterprises and they cannot provide ICT service for businesses. It seems that the future is uncertain.

© Springer Nature Singapore Pte Ltd. 2018
Z. Li, *Telecommunication 4.0*, DOI 10.1007/978-981-10-6301-5_6

1.1 Looking for Next Dividend

The growth of users was a main contributor to the fast development of operators. But this contribution is becoming less significant. Compared with the huge success of Internet giants, the development of the telecommunication industry has been lackluster. All parties in the industry are exploring next dividend. Recently, trying to make its own contribution to future development of the industry, Huawei Technologies Co. Ltd broke the development of the industry into several stages, including demographic dividend, cell phone data dividend, data dividend and information dividend.

Demographic Dividend: In the first stage of telecommunication, this technology enabled communication to transcend geographical barriers and occur at any time. In this stage, universal access to telecommunication was realized, satisfying people's basic need for communication. Operators achieved a steady growth in revenue with a growing number of users. Insufficient coverage of the network was the main challenge in this stage.

Cell phone Data Dividend: With the completion of the upgrading of 3G/4G, people's demand for data increased. Communication will shift from words to pictures and videos. In the next 5 years, the number of mobile broadband users will grow by 3–4 billion and demand for data will increase by 4–6 times. The main problem in this stage is the gap between consumers' demand for data and broadband networks.

Data dividend: Technologies such as cloud computing and big data enable companies to move their offline businesses online. Internal IT applications of businesses will be moved to the cloud for centralized supply. Core businesses of operators will expand to provide companies with cloud and big data services and these two services will be key to increasing revenue. In this stage, it is essential to ensure a good user experience.

Information Dividend: Huawei estimates that by 2025, there will be 100 billion connections around the world, creating an interconnected world. The digital world will transcend the physical world. In the digital world, various possibilities will be discovered and constant innovation means anything is possible, which leads to information dividend. In this stage, the key is to create a sound environment for innovation in order for a healthy development of the digital world.

In recent years, we can see that many operators around the globe have been taking measures to have a bigger share of dividend in the future. For example, Spain is promoting UNICA based on the cloud to improve its ICT infrastructure in order to maximize the realization of its digital assets. AT&T Inc is implementing Domain 2.0, which is aimed at making 75% of its networks use software to support its business concerning corporate clients in order to become a Supper Carrier. Softbank, in addition to its main businesses of telecommunication and internet, is expanding into the areas of new energy, robots, transportation, and application software to have a head start in the interconnected world.

1.2   Digitalization Is Indispensable to a Successful Future

Digital natives born in the 1990s will be the main consumers for all industries. Therefore, losing these natives means failure to win the future. Based on the needs of end users, the key is user experience. Huawei uses the key word ROADS to describe the era of experience. R stands for Real-time, O for On-demand, A for All-online, D for DIY and S for Social. In the industry of telecommunication and internet, user experience is becoming the core of strategy. In addition, ROADS has become the criteria for user experience for many industries. Good experience can bring about a higher premium and a 1% difference may increase the value of the network by several times (Fig. 1).

Rapid development of mobile internet, cloud computing, big data and the Internet of Things leads to a growing demand for telecommunication networks, which will contribute to explosive growth of data, connection and volume. Therefore, in the future, operators need to build and provide a very wide network. Looking at the history of telecommunication, we can see that from analog to program control, ATM to IP and 3G to 4G, what is most suitable for operators is a standardized huge market. Established telecommunication networks in which years of investment have been made are the most valuable asset for operators. Networks were, are and will be the core business of operators. Operators should build new business cooperation system by taking advantages of the heavy networks asset, open network capability and system capability.

From voice to data, businesses of operators have become diverse. In the past, 80% of the revenue came from 20% of the businesses but now 80% of the revenue comes from 60% of the businesses. Businesses in the digital age are more diverse and fragmented. The lifecycle of digital businesses is increasingly short, from several years to months or even weeks.

To satisfy the need for ROADS experience and adapt to fragmented businesses, openness and cooperation in the digital age, operators should significantly cut TTM (Time To Marketing) and the cost of trial and error. In the future, meeting new demand depends on the capability of networks. Hence, operators should make quick responses. However, the existing chimney network structure and traditional way of operation make the introduction of new business too slow and OPEX

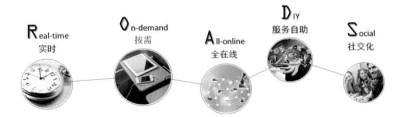

Fig. 1 ROADS reflects end users' demand for services in the era of Telecommunication 4.0

(Operating Expense) continuously high. Digitalization is essential to tackling these problems and growing the telecommunication industry.

Digitization is essential to tackling these problems and growing the communication industry. Operators should shift from a self-centered mode of "network-operation-experience (homogenous)" to a user experience-oriented mode of "(heterogeneous) experience-operation-network".

To some extent, the essence of digitization is "traditional communication + Internet" whose chief target is to improve business agility and reduce the cost of trial and error. Digitization is aimed at restructuring the production and operation systems and business processes to build a flexible network, an open and win-win business ecosystem, a responsive operational system and efficient flat organization. All of these are designed to create a flexible and open business system to broaden sources of income, cut OPEX(Operating Expense) and improve business agility.

In the future information society, information interaction is mainly dependent on videos. They will be ubiquitous throughout people's lives and used in every walk of life, becoming a new way of life and life experience. In the future, 70–90% of the data of the networks of operators will be videos and they will become a basic business for operators. Videos will include entertainment videos, industrial videos and communication videos and so on. End-to-end experience is vital to videos and the experience is closely related to networks. This requires new demand for networks because the effect of the application of networks is closely associated with user experience. Therefore, based on their advantages, operators should build a network for the best video experience to create a positive business cycle.

1.3   Devising Winning Strategies in the Era of Telecommunication 4.0

As China Mobile Communications Corporation says, the global communication industry is entering the era of 4.0, which features the integration of IT and CT. Cloud computing, virtualization, softwarization, open source and big data enables the communication industry to change everything.

To have more dividend in the era of 4.0, operators need to improve their ICT infrastructure and make their IT systems a value-producing system instead of an internal support system. Therefore, operators need to take the following measures to digitize their businesses.

1. Applying cloud computing and software into ICT infrastructure

On the one hand, cloud computing is becoming the catalyst for the combination of IT and CT. Future IT infrastructure will be based on cloud computing, significantly increasing the resource efficiency of the infrastructure. Meanwhile, SDN and NFV technologies should be used to build a software-defined network to cut TTM (Time to Market) and increase the efficiency of the operational system and better integrate ROADS into user experience.

On the other hand, cloud computing will become a very important opportunity for operators. In the future, more than 50% of companies will put over 50% of their assets in third party centers and over 60% of storage volume will be provided by cloud service providers. Compared with multinational cloud service providers, operators enjoy advantages in terms of local brand and resources. Cloud services will become a new source of revenue for these operators and a cloud-based telecommunication infrastructure is the basis for providing cloud services. Cloud computing is not only a kind of technology, but also a business model and an ecosystem.

Huawei has been working on data center solutions for a long time and making great efforts to provide global operators with distributed cloud structures and a platform for IT infrastructure. The two find wide applications in networks of operators around the world and receive recognition from a growing number of clients.

2. **Building an automatic and smart operational support system of the next generation**

The goal of this system is to cover the resources, businesses and operation of operators. It is a whole operating system Huawei calls Telco OS. It has several core modules.

First, IES (Infrastructure Enabling System) is used to improve the flexibility of the bottom layer ICT infrastructure. Second, the big data engine can help operators take full advantage of huge amounts of user data, better understand user stories and create a valuable business strategy to make operation excellent. Last, BES (Business Enabling System) improves agility and flexibility in terms of the product development and plan design of operators, the business development of partners and the purchase of products and services.

3. **Innovative business beyond connection**

Cloud-based ICT infrastructure, an operational support system featuring ROADS and a capability opening platform enable operators to provide individuals and businesses with a high quality connection and cloud service, attract all industries and third party developers to join their ecosystems, develop innovative applications for vertical industries and segmented markets, such as the Internet of Vehicles and the Internet of Things, and end user-oriented niche markets, which can enrich people's life and broaden the source of income for operators.

1.4 Working with China Mobile for Digitization

The digitizing of the communication industry has begun, with a growing number of operators around the world taking actions to digitize their businesses. At the Mobile World Congress Shanghai, held on July, 2015, we saw that China Mobile unveiled its strategic vision 2020—it is dedicated to changing life via mobile, building smart pipes and becoming a trustworthy expert on digital service. In addition, China Mobile also released its NovoNet 2020. We can see that in terms of digitization, China Mobile has a very clear strategic vision and measures.

Huawei, a strategic partner of many leading operators around the globe including China Mobile, also sees digitization as an important opportunity to change the global communication industry. As early as 2008, Huawei started to engage in the field of cloud computing, with a focus on the private cloud. In the year of 2008, it kicked off the "Cloud Sail Initiative", announcing the formal entry into the cloud computing market. Then, a series of solutions followed. It has come a long way in the market by virtue of its established strength in research and development. In 2012, it proposed the idea of SoftCOM, a strategy for the transition of future communication networks. This strategy, aimed at the digital transition of operators, covers all fields, from the overall structure, network, operating to businesses so as to help operators enter into the digital era. Meanwhile, Huawei also understands that the future transition is no less than a comprehensive and systemic project, in which services play an essential part except for products. Therefore, it has adopted a dual-wheel strategy driven by both products and services instead of the former "driven by products, supported by services". This makes it easier for customers to transit in the future by thanks to the one-stop software and hardware solutions (products, consulting, planning, integration and support) offered by Huawei.

Hard it is to forge ahead amid storms, it is worthwhile to reap the fruits thereafter. It is predictable that, the road to digital transition of the communication industry is destined to be a long and difficult journey. It is a test for both operators and providers. Only one with persistence will win. As a strategic partner which has accompanied the operators who have achieved rapid growth over the past two decades, Huawei wishes to face up to the challenges, present and future in partnership with China Mobile Communication Corporation and other partners in this field, jointly promotes the process of digital transition in this field, and embraces a better interconnected world.

2 The Ongoing Future

Werner Schaefer, Vice President of Hewlett-Packard Company

Over the past 30 years, HP China has been committed to China's communications undertakings, as it has introduced international expertise and advanced products into China and provided high-quality communications services.

Since its entry to China's market in 1985, HP has experienced a series of reforms. It has been improving service quality through increasingly upgraded product mix and advanced technologies. This year, HP is divided into two branches. The division for providing products for enterprises forms HPE, which ensures HP better integrates its advantageous resources to compete in the market of enterprise products.

The 1980s witnessed the emergence of mobile communications. In 1978, the first mobile communications system was put into operation in Chicago, ushering in the era of mobile communications analogue. People's demand at that time was merely voice services. By the mid-to-late 1980s, or the digital era, mobile communications started to carry digital services and changed people's life accordingly. In the 21st century, as people's demand for digital services and mobile connectivity grow, mobile communications embraced another turning point. The demands for network integration, service integration and for the transformation of operators pushed operators to explore new ways of network development. This situation ushered in the IP era.

If we look back at the history of the mobile communications, we will find that mobile communications is essentially driven by technological advances. This is turn in the analogue ear and the digital era. Even in the IP era, mobile communications is essentially an IP-based communications network (carriage, protocols). This progress could be regarded as the initial step of the combination of CT and IT.

According to the 36th *Statistics Report on Internet Development in China* issued by China Internet Network Information Center (CNNIC) in 2015, as of June 2015, China's cell internet users numbered 594 million (88.9% of all netizens), and the average usage time each week reached 25.6 h, as illustrated by Fig. 2.

As the way of internet usage and time change, services that can be completed online also grow exponentially. Today, we can use cell phone to enjoy music and games, read novels and news, go shopping and check E-mails anytime anywhere. Hence, it is no exaggeration to say that mobile connectivity changes every people's life, as illustrated by Table 1.

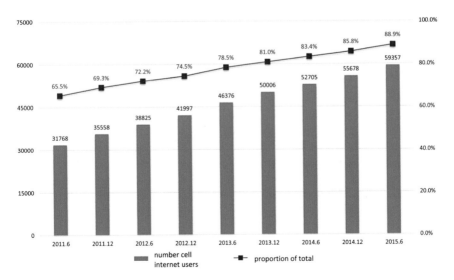

Fig. 2 Proportion of China's cell internet users to total internet users (*source statistical survey on internet development in China* from CNNIC, June 2015)

Table 1 Numbers of service users and proportion to total

Applications	Dec. 2014		June 2015		Half-year growth rate (%)
	Numbers of uses (in ten thousands)	Proportions to total (%)	Numbers of uses (in ten thousands)	Proportions to total (%)	
Instant communications	60,626	90.8	58,776	90.6	3.1
Online news	55,467	83.1	51,894	80.0	6.9
Search engine	53,615	80.3	52,223	80.5	2.7
Online music	48,046	72.0	47,807	73.7	0.5
Blog/personal zone	47,457	71.1	46,679	72.0	1.7
Online video	46,121	69.1	43,298	66.7	6.4
Online game	38,021	56.9	36,585	56.4	3.9
Online shopping	37,391	56.0	36,142	55.7	3.5
Micro blog	20,432	30.6	24,884	38.4	−17.9
Online literature	28,467	42.6	29,385	45.3	−3.1
Online payment	35,886	53.7	30,431	46.9	17.9
E-mail	24,511	36.7	25,178	38.8	−2.6
Online banking	30,696	46.0	28,214	43.5	8.8
Tourism reservation	22,903	34.3	22,173	34.2	3.3
Group buying	17,639	26.4	17,267	26.6	2.2
Forum BBS	12,007	18.0	12,908	19.9	−7.0
Online stock or fund	5628	8.4	3819	5.9	47.4
Internet financial investment	7849	11.8	7849	12.1	0.0

Although it is not easy for us to know whether the growth in demand promotes technological advances or the later stimulates the former. But one thing is for sure, today's mobile communications faces unprecedented pressure for transformation.

Operators have relied on transmission/connection services for a long time. The *Statistical Report* suggests that broadband and mobile internet have created a connected world that allows access to any service users need while extra use of the internet does not generate extra income for operators. Income decline per bit is quite obvious, and even costs of network equipment and technological advances cannot be reduced. Figure 3 illustrates the relationship between income per bit and cost per bit of operators around the globe. From 2017 to 2018, the cost is expected to be higher than the income, which may cause problems in infrastructure investment.

(Unit: US dollar)

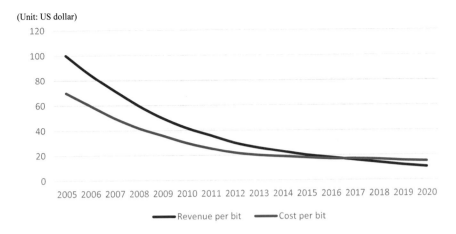

Fig. 3 The relationship between income per bit and cost per bit of operators around the globe (*Unit* US dollar)

Maybe in the near future, we will see that the historical effects from such relationship will put pressure on operators in their capital budget. Over the years, financial analysts have been talking about the long-term pressure of capital expenditures and deferred spending plans. Now, policymakers have already realized that problems will not be addressed by canceling several projects. If the income per bit keeps declining, return on infrastructure investment will fall below what the Chief Financial Officers (CFOs) expected. Operators, for higher returns, may turn to emerging markets and stop investing in their own. Mobile connectivity will probably not work out as a result of capacity issue.

Of course, nobody wants to see that happen. Operators are striving to find solutions. For SDN, the cheaper "white box" is expected to replace switches and routers so as to reduce capital expenditures. NFV is also aimed at replacing expensive network equipment. Unlike the precedents, this transformation will bring about more changes in the entire communications industry. However, there are also challenges.

The first lies in capital expenditures. New technologies will undoubtedly help reduce capital expenditures, yet making obvious changes are not easy. Many insiders have reached consensus: capital expenditures might be reduced by 15–25% if all strategies for this purpose are used; but changes in cost may not manifest till years later because installing network equipment requires long time cycle.

The second lies in changes of operation efficiency. According to investigations, half of the total equipment cost is for operation and the other is for capital expenditures. Assume that the total remains equal, a 50-percent reduction of operation costs will come with a 50-percent increase of capital expenditures. If operators improve their efficiency, their cost can be further reduced and more areas of investment will be generated. However, problem still exists. The expectation that SDN and NFV reduce operation costs has not yet been proven by quantitative and qualitative analysis. In practice, reducing the cost of a certain area is more difficult, but still a job CFOs must do.

Third, flexibility of services is also critical. Today we have already realized that flexibility of services often means maximization of profits. A popular view is about OTT service. Many experts believe that if operators are more flexible, they can stand out and remain at the top of the market. However, most successes in OTT are achieved through capitals earned from advertisements. In 2015, total expenditures on these ads may be lower than $600 billion, compared with $550 billion of telecommunications expenditures. Even if all ad expenditures are turned into online services without any discount or loss, the earnings can hardly cover capital budgets of network operators around the world.

Although opportunities and challenges coexist, we still have good reasons to be confident of the new communications era. The good news for operators is that demands in mobile services and Internet of Things (IoT) have offered new opportunities, and Content Delivery Network (CDN) is in transition from an ad-driven model into a paid-service one. In the future, users will be transferred to a virtual and knowledge-processing network. This network is based on users' locations, social frameworks and sensors around users. And "services" of the network can be regarded as "policy-making products". This means that the network can draw on all powers to process information and provide answers and practical ideas, not just "knowledge".

Although many people are talking about "transformation of business model", do communications operators really need a new model? They have already sold their services to almost every family on the earth. And they know how to utilize expensive and well-built infrastructure to create services and turn them into successful business. Operators need new infrastructure—not physical network construction to improve flows of broadband but cloud to provide services. The key is to ensure operation and flexibility of the new infrastructure thus to help quickly reach goals at costs and risks acceptable to CFOs.

The transformation entails four stages:

The first stage is decoupling. This means separating software from hardware, the goal is to ensure that virtual network can operate in standardized and open platforms.

The second stage is virtualization. Virtualization technology helps improve utilization rate of resources and workplace efficiency. Many operators have already benefited from virtualization in IT development.

The third stage is cloud application. Virtual computing, network and stored resources will be managed in a flexible pool. More effective coordination among demands for application, automated services and delivery will be allowed.

The fourth stage is division. Virtualized resources will be divided, and virtual network itself will be divided into smaller function units—usually called microservices. Microservices of operators and their clients can form new mix and create more services in less time. It is expected that reaching this stage will take years.

Ultimately, operators will secure a platform that enables them to choose partners freely and application software as needed. Evidently, what operators need is like an open ecosystem. In this large ecosystem, operators need to face the following three partners:

Technical partners—technology companies and suppliers, including suppliers of network equipment, original equipment, and telecommunications operators. They will conduct cooperation in technological innovation, integration and basic framework will be conducted.

Application partners—application developers that provide application software and functions.

Service partners—system integrators.

As a traditional IT company, HPE has not only solid technological foundation but also valuable experience in conducting business with telecommunications users. These technologies and experience provide guarantees for operators in this transformation.

1. Support from various products

As the world's largest IT enterprise, HPE is able to offer the best support for operators with its advantages in products, technologies and services of enterprise IT market, including the following:

High-quality and standardized hardware equipment, including the world's best-selling x86 blade servers (blade and rack), the best choice for virtualization and system integration.

Abundant software product lines, Among them, HPE OneView offers unified and comprehensive infrastructure management tools; HPE Director is used to organize NFV network functions and to manage and monitor global resources (cross data centers, cross resource pools).

Software-defined storage devices. HPE possesses the world's largest OpenFlow switch lines and over 20 million ports of OpenFlow switches have been installed and put into operation. Commercial SDN controllers are being intensively developed.

2. Participants and contributors of professional standards

For the currently popular NFV technologies, HPE promises that OpenNFV will use at all levels open technologies, including but not limited to compatibility of the ETSI framework, support of the OpenStack API, network technologies of the OpenFlow, TM Forum and compliance with the ITIL standards. HPE participates through investment in the standard-making process to deliver on its promise.

Besides, HPE is a key leader of the OpenStack project in which HPE has two members of the Board of Directors and three members of the Technology Committee. It is also a major contributor of the OpenStack community. HPE Helion OpenStack is being upgraded to a telecommunication version that ensures higher reliability for improving functions of the NFV loads, network, security and deployment management.

3. **HPE Open NFV Partner Program**

Telecommunication operators all hope that NFV is a platform that enables free choice of partners. Through this platform, operators are free to choose application software as needed. The ultimate NFV framework may incorporate technologies and products from various parties. In order to support such openness and standardization, HPE launches the Open NFV Partner Program. The program includes small and medium-sized providers of independent software, leading network equipment providers, various technology providers and service providers. Operators are free to choose and mix different products and technologies, and achieve smooth migration of their network to NFV based on their business and IT.

Reviewing the history of mobile communications, we can see that every transformation has brought about profound changes in the entire industry and even in every people's life. Today, it is lucky for every one of us that we are in another transformation, yet it offers us both opportunities and challenges. As IT and CT further converge, operators need to adjust the current network so that it is more closely integrated with cloud technology. Meanwhile, operators also need to curb risks in the investment they need to ensure the balance of input and output. As for equipment providers, telecommunications operators expect from them more competitive pricing, better performance and more transparent functions; promise of providing source codes, workable typesetting strategies and progressive, expandable software design so as to achieve full use of the cloud.

These may sound harsh for all providers. However, we need to bear in mind that in any major technological transformation some enterprises grow unexpectedly while some go bankrupt as a result of competition. For HPE, the accumulation in the past meets just right operators' expectations. HPE has experienced many transformations like this and developed itself each time. In this new transformation of mobile communications, HPE is still a reliable partner.

3   A New Win-Win Solution in the Telecommunication 4.0 Era

Ding Lei, Founder, Chairman & CEO of Netease

3.1   What Will Happen in Our Industry When CT and IT Are Fully Integrated

Over the years, telcommunication technologies have evolved from 1.0, 2.0 to 3.0. Meanwhile, in China, the large-scale internet infrastructure invested by telecommunication companies helped to give birth to three major web portal sites such as Sina, NetEase, Sohu and internet giants such as Baidu, Alibaba and Tencent. In earlier years, IT companies made huge profits by providing internet services for users through networks built by telecommunication companies, while the latter got their share mainly by charging traffic fees. Therefore, in this period, the major role for telecommunication companies was data transmission channels. However, as we step into the era of Mobile-Internet, the proportion of mobile traffic has increased with the popularity of 4G networks. Telecommunication companies have become less perceptible to users who are more familiar with mobile applications developed by IT companies. Moreover, some social application companies (for example, WeChat) have introduced new features including textual messages and audio messages into their products. This has posed a threat to the core line of business for telecommunication companies whose main source of revenue has, consequently, turned into traffic charges. As a result, there has been a growing tendency of channelization among telecommunication companies.

Today, as the integration of IT (Information Technology) and CT (Communication Technology) goes deeper, the industry model is also transforming. A new landscape is taking shape in our industry, involving both IT companies and telecommunication companies. Cloud Computing, Big Data, Internet Plus, are all products of the ongoing integration. With the emergence of new technologies, new products and new services, the whole industry as well as its value chain will be restructured. Following are some main tendencies.

1. Telecommunication Companies: A Major Shift from IP-Based Networks to IT-Based Networks

Previously, networks built by telecommunication companies were IP-based and traffic-oriented, which means they were mainly intended for data transmission. However, in the era of Telecommunication 4.0, these networks will be IT-based and application-oriented, thus requiring the support of SDN (Software-Defined Networking), NFV (Network Function Virtualization) and Automation. SDN separates the control plane from the forwarding plane so that forwarding devices are standardized, and application programming interfaces are programmable. In this way, applications will able to control network forwarding based on their needs. By exercising centralized control, carriers can improve management efficiency and lower management costs, Their networks, as a whole, can be further softwarized, generalized and centralized. NFV technologies enable building network functions on general-purpose hardware instead of special-purpose hardware. NFV technologies fall into three categories, namely, virtual devices, virtual services and virtual channels. Virtual devices can build multiple network devices in a single computational node using virtualization technology, for example, virtual routers and virtual firewalls. Virtual services can integrate a group of servers into a single device to provide internet services such as load balancing services and traffic cleaning services. Virtual channels can turn connection resources such as private lines into multiple channels and provide virtual access for devices using virtualization technologies. They can also work with SDN technologies to realize dynamic scheduling, thereby improving resource efficiency. What's more, NFV can turn general-purpose hardware into a vast pool of shared resources to provide better, cheaper services for IT companies. In addition, telecommunication companies need to provide automatic programming tools and make full use of these resources to ensure quick responses, to expand and upgrade, and to improve their operation and maintenance efficiency.

2. Traditional IT & CT: All-Round Internetiolization

Traditional information technologies provide technical support for companies, in other words, what is going on inside these companies are the main concerns. However, with the rapid development of Cloud Computing, information technologies have been redefined by the quick response, highly-personalized features and flexible services made possible by Cloud Infrastructure, Cloud Platform and Cloud Service. Today, companies are able to leave emails, backups as well as other communication and security issues to the Cloud System and focus more on their own business.

Faced with the challenge of OTT (Over the Top) services and the ever-growing tendency of channelization, in view of long-term benefits, telecommunication companies have no choice but to move faster to embrace the upcoming internetiolization. Firstly, they should provide high-security, high-quality networks for the IT industry, for example, solutions customized for mobile applications and video applications, and explore win-win strategies with their business partners. Secondly, they should build their networks in a planned way using intelligent technologies. Many telecommunication companies used to connect IDC (inter-data-center) networks to intercity networks instead of building exclusive networks for IT companies. Therefore, in order to meet their extra needs, some IT giants have to do this by themselves. Today, with the fast expansion of data centers, the traffic generated by IDC have approached or even exceeded that by inter-city networks. To remain attractive to customers, the first thing those telecommunication companies should do is to manage their channels more smartly to better serve their clients.

3. Cloud Computing and Internet Plus: Greater Influence on the Future of IT Industry

As a revolutionary technology in the IT industry, traditional Cloud Computing provides a cheaper and more flexible way to process and save data. However, while this technology strives to offer high-quality IT resources and strong technical support, it pays less attention to user friendliness and product experience, which can result in rather complicated solutions. Judging from the history of the IT industry, we (NetEase) think the emphasis of Cloud Computing will shift from IT resource supporting to IT services sustaining. A new type of Cloud Computing will aim at providing a series of technical support to help clients for their business innovations. This will happen first in the core segments of a company, involving every stage from developing and testing to launching and operating, and then, in other departments, including production, human resources, finance and law. The most popular products will be systemic support in communicating, testing and operating rather than pure IT resource support in computation, data storage and network connection. The new Cloud Computing will focus more on user friendliness and product experience which lead to light and agile solutions with further competitiveness Accordingly, its services will be more personalized instead of purely technical. This trend will challenge the whole industry, placing higher demand on the performance of communication infrastructures. Cloud Computing providers and clients will require more flexible, personalized services to satisfy the integration needs of different enterprises and industries. And traditional industries will, inevitably, be fully integrated with the Internet, a new concept known as "Internet Plus". The application of Internet Plus will enable the integration of private IT infrastructures, public IT cloud facilities and user terminals, which also demand better performance, more personalized service and higher security from the communication infrastructure.

4. Telecom Carriers and Internet Companies: Pursuing Closer Win-Win Cooperation through Internet Plus

As two indispensable parts of the IT industry, internet companies and telecom carriers both have their own advantages. On one hand, the vast resources owned by the former, including wide coverage of sales networks, large construction and maintenance workforces, huge accumulation of users and well-maintained government relations are attractive to the latter. On the other hand, the innovative thinking, the advanced technologies as well as the capacity of quick responses of the latter are also desirable to the former. With the rapid development of Cloud Computing and Internet Plus, those two kinds of companies can work together in many areas such as Cloud Computing and OTT technologies to explore new business models of win-win cooperation.

As an internet company, NetEase also hopes that telecom carriers can transform themselves from resource providers to resource operators. By managing data channels and applying Cloud Computing technologies, carriers can help clients better apply the Internet Plus strategy. In terms of data channels, they can, with the help of technology, provide more flexible, secure and reliable services for internet companies like NetEase. Let me take private lines as an example. Now the cost of private lines between data centers are very high; in order to ensure quality of service and disaster tolerance, clients tend to rent spare bandwidth on these lines and deploy their own network access devices on both ends. But as the resources of private lines are rather limited, if clients have temporary or regular needs for massive data transmission, they will have to bulid VPNs themselves and transfer data through public networks. In this way, the quality of service will be compromised. When the bandwidth of private lines can no longer support the growth of traffic, it will take a long time for carriers to broaden or upgrade it, which can even have a negative influence on client's business.

If telecom carriers unify the management of DCI private lines using SDN technologies to provide connection resources in a virtualized way as well as flexible internet access solutions based on NFV technologies, clients will be able to use on-demand private lines more freely and conveniently. For telecom carriers, this practice can not only meet the needs of internet companies, but also improve the utilization of private lines, thereby lowering material and management costs. This is quite similar to the "Cloud Computizing", that is to say, to provide services in the same way as a Cloud System. Such strategies and ways of thinking suggest a new model of cooperation between telecom carriers and internet companies that will make room for future development.

3.2 NetEase: Our Plan and Technical Preparation for Telecommunication 4.0 Era

Though NFV and SDN technologies haven't been around for a long time, they both develop rapidly and represent the future of internet technologies. Our R&D departments have already begun to explore and apply both of them as early as three years ago within data centers when building our new Cloud Computing infrastructure platform.

As for NFV technologies, NetEase Cloud Comb (a public Cloud System) independently developed NFV cloud services including traffic cleaning, load balancing and virtual private networks based on open-source technologies such as OSPF (Open Shortest Path First) protocol, LVS (Linux Virtual Server), Nginx and Intel DPDK (Data Plane Development Kit). As for SDN technologies, NetEase Cloud Computing independently developed an efficient and flexible SDN network controller based on open-source technologies such as OpenStack, OpenFlow and OpenvSwitch (Open Virtual Switch) and VxLan (Virtual Extensible LAN) to meet our own needs in network building. In this way, we have realized secure communication between networks inside and outside of the Cloud System, safe isolation of Cloud Computing users and flexible configuration of the Cloud Network.

NetEase has profited a lot from the exploration and application of NFV and SDN technologies. Firstly, by applying NFV technologies, we substituted high-cost, special-purpose network devices, for example, traffic cleaning devices, for low-cost, general-purpose ones, thus significantly lowering the hardware cost for our company. In this way, our basic IT system became much more extendable so as to facilitate the company's future development. Secondly, by applying the latest SDN technologies, we not only improved the network security for the Cloud Computing platform, but also simplified network configurations, thus lowering the human cost of operation and maintenance. Last and foremost, in doing all this, we are able to keep up with the latest trends in network technologies and, to gain more experience so as to lay a solid foundation for future development.

Although NetEase already has some achievements in the application of NFV and SDN technologies, both of them, as we see, are still very immature, with a lot of improvements to be desired. This is also where we are heading in the next stage.

By replacing special-purpose hardware with software and general-purpose servers, NFV technologies will improve flexibility, but at the same time, almost inevitably, lower performance. Therefore, general-purpose hardware producers and internet companies should work together to improve and optimize hardware performance.

Today, the widely used applications of SDN technologies are still restricted within data centers. However, with the rapid development of Cloud Computing, the needs for the realization of traffic scheduling and cross-room IP floating in WAN (Wide Area Network) using SDN technologies will continue to grow.

With the diversification of user needs, traditional special-purpose hardware will no longer compatible with the rapid changing situation. The NFV-applying for special-purpose devices is a new area that needs further exploring.

3.3 Conclusion

In the era of Telecommunication 4.0, when telecom carriers transform themselves into resource operators, the ICT industry will also change to keep up with the technology trends represented by Cloud Computing and Internet Plus. In the Telecommunication 1.0, 2.0 and 3.0 eras, the division of labor within the industry chain was very clear. Hardware producers provided special-purpose network devices for telecom carriers, who used them to build IP-based networks to provide traffic channel services for internet companies and terminal users. Today, as the integration of IT and CT goes deeper, a new landscape is taking shape in our industry, involving telecom carriers, telecommunication companies, IT companies and internet companies. IT companies will provide general-purpose software, hardware and systems, while telecommunication companies will provide interface-programmable network devices. Telecom carriers will IT-ize themselves by applying NFV and SDN technologies on general-purpose hardware and network forwarding devices. In doing this, they can realize smart network management, lower management costs and provide more flexible and customized services. Internet companies will provide technical support and provide innovative internet services. In the era of Telecommunication, with the emergence of new technologies, new products and new services, the whole industry as well as its value chain will be restructured.

Following the Telecommunication 4.0 trend, internet companies and telecom carriers will be increasingly interdependent and desirous of win-win cooperation. For example, internet companies like NetEase have a lot of technological accumulation in Cloud Computing which enables them to provide responsive services as well as support in basic technologies and solutions. Telecom carriers, on the other hand, have their own traditional advantages such as well-established brands, abundant IDC network resources as well as enormous channels and users. In view of this, they can work in close partnership, making the best of both sides to realize deep integration of IT and CT, to change the whole industry with the help of Cloud Computing technologies and network connections. Such cooperation model may be extended to other areas, such as Big Data and Internet Plus, to promote win-win cooperation in the whole society.

4   Intelligent Communications 4.0 in the New Era

Mr. Edward Suning Tian

Chairman of China Broadband Capital Partners, Executive Chairman of AsiaInfo

In the era of industrial economy, enterprise-customer relationships were mainly represented at the process of products purchase and customer services. However, most enterprises were far from their customers; or to put it another way, they were not well acquainted with their target customers, the state of customer experience and next-time purchase demand. With advertising being the major marketing approach and multi-layered purchase behavior, after-sale service presented itself as a weak point. Customer service centers just dealt with problems that had occurred, having no idea of customers' use procedure. All of these aggravated the contradiction between mass and standardized production and increasingly-growing personalized demand.

With the decreasing cost of computing, ubiquitous networks and the popularization of low-cost smart terminals in particular, the information economy has shown a character of feedback economy. Enterprises boasting superb products and services are embracing brand-new opportunities right through product management, services and customer operating, ushering in the era of customer operator.

The term of operator refers to telecom operators. They can be reached and help to solve problems around the clock. The call number of customer service such as 114, 10086, and 10010 is one typical aspect of operators. "Always online, instant response" is the capability owned by operators. Despite a number of challenges posed by new technologies and applications from the market, for operators, Telecommunication 4.0 is the way moving forward and a typical representative of strong customer relationship. Personal identity, data traffic, payment bound by telecommunication services still constitutes an essential infrastructure in our lives. Operators also signify an effective business model, in which customers are charged based on quality, distance and usage, acting as a spur to the almost ubiquitous telecommunications network established by the industry within a century. Such paramount scientific and technological innovations as semi-conductors, Unix operating systems, wireless communication, optical fiber technology all emerged in the ecological system of telecom operators. It can be said that telecom operators are customer operators with strong customer relationships and frequently used applications in the era of industrial economy.

Entering the era of networked industries, In order to become customer-oriented and "SUPER" operators, telecom operators need to transform its BOSS (Business & Operation Support System). "SUPER" means capabilities of Social, U-centric, Platform, and Exchange center in Real-time, as shown in Fig. 4.

Smart terminals and the Internet of Things are making computing more socialized. Just imagine: our daily physiological data are recorded by wristbands and watches; room, air-conditioner, washing machine, refrigerator and others can

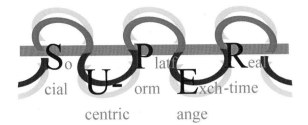

Fig. 4 Supper operators

record their state of usage. Cars can record, count and analyze locations and usage, while powerful cloud computing capabilities are backstage processing, analyzing and giving feedback to such devices, applications and products. These essential factors, added up, make it possible for all future enterprises to become competent in running customer-oriented operation and thus changing business model.

In the era of industrial economy, a refrigerator manufacturing enterprise was seldom related to their customers after selling products. It did not know who was using the refrigerator at what place, let alone how the refrigerator was being used and what kinds of food were stored. It was an enterprise producing standardized products, with little knowledge of their customers. Customer service centers only solved problems when customers complained. Picture this, in the era of networked industries, refrigerator producers would mark every refrigerator and access to such data as state of usage, food stored, and frequency of usage through Internet of Things chips and intelligent images. In this condition, business model of refrigerators would also face a change. It may become a window of knowing your food preferences and a gateway to food e-business; while producers can rent refrigerator to you, providing food, predicting food that you might need, thus becoming your food operator, as shown in Fig. 5.

In the era of networked industries, the decisive factor for enterprises to succeed is the transform of concept, technology and business model. Instead of being producers and service providers, they shall become 24-hour online customer-oriented operators, experts and predictors of customers' demand by virtue of products and services that strengthen relationships with customers.

Lodging Service Operator   Trip Service Operator   From Refrigerator Producer to Food Service Operator

In the era of Industry Internet , the key for success is to build strong relationship by providing product and service and to become the custom operator.

Fig. 5 "Operator model" in the era of networked industries

In the era of networked industries, the key for enterprises to become successful lies in becoming "customer-oriented operators" by building "strong relationships" with customers through products and services.

Communications 4.0 has laid the technological and network foundation for customer-oriented operators. A new generation of operation support system is needed for a telecom operator to achieve such transition, which links, manages, and operates billions of customers, tens of billions of smart devices and even more mobile applications. This is a challenge to information architecture, database software, application systems, as well as a business opportunity for innovation. AsiaInfo's next generation BOSS system empowers telecommunication operators with SUPER capability, thus helping companies become customer-oriented operators. It enables enterprises to establish customer-oriented unified processes, transform all capabilities of products, identity, orders management, charging and clearing offered by traditional BOSS system into services for open use, and realize "Operating Everything".

AsiaInfo has always been an enabler for telecom operators in the past two decades. In the era of communications 4.0, AsiaInfo will provide the new generation of BOSS system to embrace the era of customer-oriented operators.

5 Intelligent Robots Lead the Fourth Industrial Revolution

Bill Huang, Founder of CloudMinds

In today's society, people continue to argue about how many industrial revolutions have occurred throughout human history. I think it is very easy to make it clear: only when several technologies have fundamentally changed both society and productivity, and when the collection of industrial innovations has increased productivity by over 100 times, can this period be called an industrial revolution; otherwise, it can only be described as industrial evolution or progress.

According to this definition, so far human history has experienced three industrial revolutions: the Machinery Industry Revolution (1750–1850), the Electric Industry Revolution (1850-1950) and the Information Industry Revolution (1950–2050) and it is quite interesting to see that each industrial revolution lasts about 100 years. During the First Industrial Revolution, the core technology products included steam engines, internal combustion engines, trains, automobiles, railway networks and highways. It enabled the power revolution from animals to machines and people were able to carry and transport more objects at a time. During the Second Industrial Revolution, the core technology products included electric lights, telegraphs, telephones, electric generators, electric motors and electricity supply networks, and thus people were able to transmit energy and messages quickly. During the Third Industrial Revolution, the core technology products include computers, Internet and computer technologies like artificial intelligence, and machines will finally replace human brains by bring powerful capabilities to calculate data and process information. Internet combines computer technology and communications technology while mobile communication combines wireless technology and computer technology, and mobile Internet has been a perfect combination of the above three. At present, the world is still in the exploration of the Third Industrial Revolution, and the tidal waves of the Internet, mobile Internet and Internet of Things all belong to this period.

Now, let's see if robots fit into industrial revolution. The robots here refer to the intelligent ones, the intelligent machines that can partly or entirely replace human roles both in functions and abilities. Thus, according to this definition, I think the next industrial revolution will be the Intelligent Robot Revolution when robots can partly or entirely replace human roles. In essence, it is humans that manufacture mechanical clones of themselves to partly or entirely replace humans for the works of autonomy and intelligence. Distinct from the three industrial revolutions previously occurring in human history which only enhanced human abilities, this industrial revolution will complete the course of the machine substitution of humans from "partial substitution" to "full substitution" in aspects of space, time, event, and scenario. For example, if an automatically driven car is created, drivers can be replaced; if a family nanny robot is made, it can help humans to take care of children and do housework; if flexible robots can be used on industrial production lines, the current industrial workers can be replaced.

By extrapolation, accounting for the constant updating of intelligent robots, disruptive changes will be ushered into all aspects in the future of human society. After the robot revolution is completed, on one hand, the jobs that rely on mechanical labor, simple operation and heavy manual labor will no longer exist, and there will be no ordinary human workers in the world; on the other hand, each of our families will be equipped with 1–2 robots to help accomplish various nanny jobs in the future. When that time arrives we will have officially entered the era of personal robotics.

If you follow this logic, the next industrial revolution, the Intelligent Robot Revolution, or more precisely labeled as the information revolution of machines, is actually a continuation of the Third Industrial Revolution but is distinguished by a huge leap in productivity.

In my opinion, the technology threshold of the robot revolution has arrived. Breakthroughs in artificial intelligence, mobile communication and battery energy storage also indicate that the era of robotics has come. Firstly, the "deep learning" research on cloud computing platforms in recent years has made several breakthroughs; secondly, 4G and the future 5G can provide unprecedented network performance and capability; thirdly, the latest Lithium battery technology has increased energy storage density to an unprecedented degree.

So, let's look at the three thresholds.

Firstly, algorithms of big data and cloud computing that can approach or even exceed human brains in deep learning have been developed. For instance, progress has been made in aspects including natural language processing, image processing, vision and even understanding of semantics and languages. Taking all this into account, I regard it as a threshold.

Secondly, mobile communication has got enough bandwidth and enough low-delay all of which can make machines perform similarly to the human brain. There will be a delay of about 30 ms when messages are conveyed to each organ in the human body, thus the "brain" can absolutely be put into cloud in this circumstance. The reaction time of controlling a machine is almost the same as that of a natural person. The appearance of this threshold is attributable to mobile communication—the world's largest 4G network has brought hope to us.

Thirdly, the miniaturization of machines and energy storage has become an inevitable trend. We can produce a 100-kilogram machine which may work 8 h, even it does not seem very realistic in these days. The Tesla electric cars have brought us great hope, but suppose we produce a 100-kilogram robot, can it really help us?

Therefore, these three thresholds have urged me to establish a company to make cloud-based robots and intelligent machines that are useful and can improve productivity.

There are four examples that can illustrate the progress in technology: First, the maturity of the autopilot technology for cars. According to Google's latest data, Google's self-driving cars have completed more than one million miles of autopilot driving on real roads since 2006. Recently, Google's research on autonomous vehicles has evolved from "road test" to "computer emulation test," which means that humans have already developed technology to completely replace human drivers.

Second, the functional demonstration of US military drones. Although being defined as remotely controlled aircraft, the drones can take their own actions even if the remote control staff issue no instructions. They can autonomously control height, evade enemy detection and keep in a hidden state when carrying out tasks in the war zone.

Third, the IBM super robot named Watson, a question-answering champion in the famous TV contest "Jeopardy!". As a robot designed to answer questions using natural spoken language, it has the capabilities of autonomous learning and autonomous reasoning. Watson became famous by competing in the well-known American TV contest "Jeopardy!" in which it defeated two of the show's previous long-standing champions. Of course, Watson is more than a question-answering robot. In May, 2015, it was adopted in 14 famous cancer centers in North America, becoming the world's first artificial-intelligence expert in cancer. By comparing treatment histories, genetic data, scans and symptoms of patients, Watson can find personalized treatments for every cancer patient, which I believe is more effective than a diagnosis given simply based on medical knowledge.

Fourth, the technological breakthrough of perception systems. In recent years, several well-known Internet companies, including Google, Facebook, iFLYTEK and Baidu, used "deep learning" technology to forge algorithms to recognize both speech and images. Some features of these new algorithms and techniques have exceeded that of the human brain.

The current demonds of social applications has reached a critical value. Firstly, looking from a contemporary perspective, along with the rapid development of industrial robots and a relatively mature supply chain, a continued increase in global labor costs and the serious issue of an aging-population create an enormous demand in society. Taking a look at the development of robots as a whole, we have also reached a critical point where the direction is going "from industry to families and individuals", and the potential to expand robots into personal and domestic services can now be realized. Secondly, looking from a practical economics perspective, anything that can rapidly increase productivity, cut costs and reduce needs for human intervention will surely be sought after by capital. Driven by profit, the capital will be invested into the production of machines that can help improve productivity. Therefore, the future market needs of intelligent and autonomous robots look optimistic.

However, an additional point needs to be addressed here. Although we say that a critical state has been reached, the entire world is still in the early exploration stage on some key issues, such as what intelligent robots should be, whether they have industrial standards and how to establish the related industrial chains. The real sense of the first generation of intelligent robots has not yet emerged.

5.1 The Intelligent Service Robot Is the Largest Market

In June 2014, the GSM Association declared that the global amount of mobile communication terminals had exceeded the global population (6.5 billion) and this number would reach 7 billion during 2015. This is the first time when a kind of electronic equipment has been possessed and used by so many people at the same time in human history. The total amount of mobile terminals will continue to increase and may reach over 10 billion.

Masayoshi Son, CEO of SoftBank, predicts that by 2040, the number of robots will have exceeded that of humans, reaching 10 billion. These robots will all be intelligent robots, and according to the research, they will look more like humans.

According to the current development rate, I believe the next 5–10 years in the cloud-based humanoid intelligent robot industry will give birth to a number of revolutionary products. Three categories will be the earliest to enter the commercial market: autonomous vehicles, household cleaning robots and patrol robots. The reasons are that they are vertically functional, the technical requirements are relatively low and the products are easier to make. In addition, China houses a huge number of families, so it is possible to quickly form a sizable market. The most extensive area of application will be home service robots, commonly known as "nanny robots". However, this type of robot is contained by the requirements for high level of intelligence and mobility. Therefore, it may cost 10 years or more before the "nanny robots" become commercially available.

The "nanny robots" usually look like humans, but actually robots can have many different shapes. There are two important reasons that we say human-shaped robots are the most significant ones. One is the psychological reason. In the future, it should be natural for us when talking to intelligent robots, but can we communicate so well with a pillar, a car or a box? It is difficult for us to accept that or establish close and friendly communication relationships with objects in nonhuman forms, so we will produce human-shaped robots in the future. The other one is environmental reason. Human-shaped robots can serve humans in the already built living environment, such as using our tools, opening doors and climbing stairs. Of course, we must design robots that can really help people instead of ones that only focus on appearances with the purpose of rapidly capturing market opportunities. Ultimately, human-shaped robots are humans' best expectations for the new generation of intelligent robots. But the key question is whether we can make them and when we can make them.

5.2 Cloud-Based Intelligent Robots Will Lead the Technological Development of Intelligent Robots

In 2010, I communicated with a professor from the Department of Bioengineering in Stanford University, and he suggested several opinions that were quite shocking: a human brain has about 100 billion neurons, and each neuron has over 500 links on average; a human brain weighs about 1500 grams, but its power consumption is only about 40 watts. He added that to clone such a brain with semiconductor technology needs a chip which is 1 million times as big as that of a human brain and weighs 2000 tons, and the power consumption may reach 27 megawatts. That is to say, if robots want to be as clever as humans, they will definitely be unable to walk around since their heads will be too heavy. However, it is quite interesting that the human brain operates based on ion mobility and electrochemistry and its speed is 1 million times slower than the moving speed of the computer's all-electronic circuit. In other words, the operating speed of the human brain is far slower than that of the computer.

From the current perspective, the weight and power consumption between the human brain and the computer differ by a factor of 1 million, which means that no matter whether future development follows Moore's Law or Post-Moore's Law, we will not see robots carrying normal-weight heads in the next 3 decades. So how can we deal with the problem of the huge gap between human technology and nature? The only solution is that we install the robot brain within the cloud, apply the mobile communication network as its nervous system and make the robot body become an "Avatar" as depicted in the movie of the same name. Though it has no head, it can be controlled by the cloud through the network. The only way to produce intelligent robots that can move is through mobile communication. So cloud computing, mobile communication and the robot "Avatar" are characteristics that I imagine when I think about the next generation of intelligent robots. The above architecture—the cloud-based intelligent robot—is the only method to realize this machine intelligence in the next 30 years.

Why do we believe that it is possible to realize this machine intelligence in the next 10 or 20 years? It is mainly because humans have already made great progress in the artificial intelligence industry. From the perspective of biology, we may not understand the real operating mechanism of the human brain, thus many nations have successively launched the "Brain Plan" which aims to study the human brain. However, we are quite clear about the results and general procedures that the brain needs to deal with, which means that there is no need for artificial intelligence to exactly mimic the operating mechanism of the human brain.

How many wireless connections will there have been by 2020? Different consulting companies have different answers and the maximum can reach 50 billion. Suppose there are 50 billion wireless connections, mobile communication and Wi-Fi are evenly split fifty-fifty, then this is already a very shockingly large scale. Currently there are only 7.5 billion global mobile communication users. But in June 2014, the number of global mobile communication users had surpassed that of the world's population, which itself is epochal. Before this, the number of pieces of electronic equipment and intelligent machines had never exceeded that of the world's population. Humans have made great breakthroughs in this area. By 2020, this number is very likely to be several times bigger than human population and this is the so-called robotics network.

In my opinion, the constraints to the emergence of the first generation of cloud-based humanoid intelligent robots are four-fold. The first is technology-intensive. On the one hand, the robots' cognition system falls in the category of high-density technology. It represents the developmental direction of technologies that are so new, so complicated and even not fully understood, such as cloud computing and artificial intelligence. It is not hard to imagine the difficulties of research and development in such field. On the other hand, the entire hardware field of robots still awaits major technological advances. For example, robots can only reach 1/10th of the level of humans by comparing the driving force per kilogram of muscle. It is necessary to develop a more lightweight synthetic material for robot body. The second is high cost. Due to currently small productivity of robots, costs for their motor and sensing systems are fairly high. Even for industrial robots, the current annual output of large multinational companies can only maintain the level of tens or hundreds of thousands units. When the market is fully opened and the productivity rises to several millions and even tens of millions units, production cost will decrease rapidly. The third is high energy consumption. Robots' operating system consumes an excessive amount of energy, which directly affects its mobility. How to move and operate, powered by a battery, for extended periods of time still remains an area that needs to be further explored. The fourth is the issue that industry standards have not yet been established. GIndustry logic shows that standards usually follows technology in general.

Now it comes back to the designing of the machine again. The machines we design must comply with humanity and provide optimal user experiences, but the intelligent robots do not have to look like humans. If they can effectively replace human roles then I think whatever designs are acceptable. However, I hold the opinion that the robots model sold out by billions are definitely going to be those that resemble humans. Selling out so many robots is as meaningful as the "Model T" car produced by Ford Motor Company. The "Model T" of robots is certainly the one built in the human form, because the machine that can ultimately replace us is the one that looks like us. Humans have designed the environment, tools and everything else for themselves since their birth. If we want to produce a robot to replace us to do all the work, it must look like us. Therefore, after whoever can produce the "Model T" of human-shaped robots, there will be no need to produce robots of other shapes. The famous robot Data in the great scientific fiction *Star Trek: The Next Generation* is a human-shaped robot.

5.3 Cloud-Based Intelligent Robots Are the Killer Application of "Ambient Intelligence" in Telecommunication 4.0

We have seen that the development of the robot industry is the core of next industrial revolution. With the world's population gradually aging, we will have a great demand for nursing staff, but we will also face the most severe shortage of human resources at the same time. The only effective solution to meet these needs is to provide sufficient human-shaped intelligent robots to help humans. No matter in house, senior service center or hospital, these intelligent robots can do all kinds of service work such as taking care of the elderly, accompanying children, cleaning and cooking. This requires us to design intelligent robots with human dexterity and skills: the extremely quick ability to react, strong ability to control, smooth movement and braking, and even expressions and emotions that can communicate with humans.

To produce such intelligent robots, we need powerful artificial intelligence and excellent robot bodies, and most importantly, we need a high-performance mobile communication system to serve as the neural network. Since the robot body is just like the human body, mobile communication network has particular significance in such application architecture. Besides providing high performance, it also needs to guarantee quality and reliability of communication as well as the security of information.

Most of the behaviors in human body are controlled by central nervous system, which is cooperated by peripheral nervous system, and our mechanical movement and sensing functions are controlled by the somatic nervous system. The control signals of our neural network transmit step by step in the form of pulses along our nerve fibers. In the peripheral nerve, these signals transmit at a velocity of 0.5–120 m/s and the conduction velocity of limb nerves is usually 60 m/s. According to this velocity, we can roughly estimate that the average transmission delay from head to foot of a 1.8-meter person is about 30 ms and the shortest delay time may still be as long as 15 ms.

Considering that the heads and bodies of cloud-based intelligent robots are separated by the mobile communication network, these parameters thus become specific requirements for the signal transmission for cloud-based intelligent robots system. Apparently, we need a very reliable and low—delay mobile communication network. Meanwhile, we should also guarantee that there is enough bandwidth to transmit all kinds of control signals, including senses of sight, hearing, touch and consciousness as well as output signals such as movement control, system control and signal display. With the technology of compression and signal handling, we only need a bandwidth of about 30 M–50 M. Of course, if robots are widely used, the density of this bandwidth may approach the density of humans per square kilometer. Therefore, signal transmission has a very high demand for the total bandwidth density.

The Communication 3.0 network can usually provide the delay performance within 50 ms, while the average velocity is of 10 M–20 M and the average bandwidth density is not high either. Thus, performance of the Communication 3.0 network is not enough to provide large numbers of high-performance cloud-based intelligent robots and we must use Telecommunication 4.0 network to guarantee the performance and services. Considering the fact that there may be 1 billion or even 10 billion cloud-based intelligent robots in the future, we can basically come to the conclusion that cloud-based intelligent robots will be the killer application of "ambient intelligence" in Telecommunication 4.0 services in the future.

To effectively serve cloud-based intelligent robots, we must establish a complete set of cloud-based robot platforms and architecture. At the same time, this system we craft can also provide services for enterprise informatization and other long-distance cloud-based mobile applications. In fact, the intelligent terminal that can communicate and realized by AI technology is an informational robot. We can equip all kinds of devices with cloud-based intelligence and they will become intelligent robots that can provide various vertical services. With the continuous advancement of technology, we can start producing some human-shaped intelligent service robots. It is an ideal that we can provide a lovely family nanny robot for each reader by 2025.

Conclusion—The Future Has Come

For years, China Mobile Communications Corporation has always been a firm supporter and practitioner of technological innovation by vigorously carrying out new technologies, optimizing and innovating network and business, and actively championing cooperation and win-win results among industries. In the era of communications 2.0, it pioneered the introduction of GSM digital telecommunications, ushering in the major development of mobile telecommunications in China; in the era of communications 3.0, it fully introduced IP technology, deploying softswitch, IMS, replacing legacy communications protocol with IP which based on DNS, and breaking the castle pattern of mobile networks; in the era of communications 4.0 where virtualization, cloud computing and other technologies are forging ahead in full steam, as a pioneer in the industry, China Mobile once again stands at the crossroad of reforming and development, follows the technological road which leveraging IT and resource pool technologies to make a flexible and general purpose mobile network. The NovoNet concept raised by China Mobile is a full interpretation of communications 4.0, as well as a bold attempt towards communications 4.0.

It is China Mobile's wish that NovoNet will reduce cost and increase effectiveness by softwarization of network function, and enhance value through the openness of networks. NovoNet, as the fundamental network in the future, will bolster China Mobile's "Three Curves" strategic development, which is to slow down the sloping trend of voice volume, promote the data traffic business development, and fully serve digitalized service, expanding new areas, thus fulfilling the demand of "Internet plus" and the Internet of Things for the communications network.

Idea and Vision

"Novo" is the root of "innovation" in Latin, which expresses China Mobile's keen interest in sustained innovation and introduction of advanced technology and its expectation for the future development of the communications network.

© Springer Nature Singapore Pte Ltd. 2018
Z. Li, *Telecommunication 4.0*, DOI 10.1007/978-981-10-6301-5

Entering the era of communications 4.0, it is necessary to carry out network transformations to achieve the goal of responding to rapid market development and satisfying customer's multi-layered demands for communications network. The idea of developing the next-generation innovative network, i.e. NovoNet, put up by China Mobile Communications Corporation in early 2015 aims to fully integrate new IT technologies, and build the next-generation network featuring "holistic resources deployment", "fully open capability", "elastic scaling" and "flexible framework adjustment". China Mobile forecasts that NovoNet will be deployed and applied in the mobile communications network, IP carrier network, transport network, data center and other fields through a planned method step by step.

China Mobile aims to achieve the following: network functions softwarization, sharing virtual resources and hardware infrastructure universalization while being capable of network programmable, control and transmission decoupling and network functions orchestration.

The vision of NovoNet is to continuously optimize network and service in such aspects as network architecture, operations management, and network openness through the two key technologies: NFV and SDN, thus making the network become programmable, deployment more flexible, dispatching more efficient, network smarter, service more open and costs lower.

To ensure the high quality of the communications network, China Mobile will also adhere to the four core appeals in realizing the vision.

1. Decoupling of hardware and software

China Mobile will unswervingly press ahead with the decoupling of hardware and software, especially realizing the decoupling of virtual network function, virtualization layer and hardware layer, putting softwarization into full play. In addition, the hardware platform, based on x86 general purpose equipment, should be high-performance, low-power consumption and high-reliability, so as to bring about the real advantages of resources pooling.

2. Carrier class quality

NovoNet will carry forward and maintain the high quality of the mobile communications network, that is, high QoS, high-reliability and high-security. High quality is the core value of the mobile communications network. Since the network extensively adopts virtualization and softwarization technologies in the future, demand for high quality will extend from purely single-device and single-system to quality guarantee of the business itself and the whole network. Besides, low-cost IT innovative technology ensures operator-level business.

3. Growing the industry through further open source

NovoNet will introduce more partners in the industry and greatly lower the difficulties in product development and integration of all parties through cooperation in open source. Without open source, there will be no openness for the industry, without which a brand-new benign industrial chain will be difficult to form and NovoNet's core idea of open and win-win results will fail to be realized.

4. Innovation of capability and business is the future

NovoNet not only represents a change of the network's own form, but also offers an innovative basis oriented toward the future and opens the elementary ability of the communications network to the third party in a more convenient way, thus making room for the third party to re-innovate. The mutual promotion of openness and innovation will bring us a miraculous future network.[1]

Vision of the Target Network

During the technology evolvement the mobile communications network has transformed from closed to open, from dedicated to general purpose. Due to the high threshold the communications network equipments used to be provided by only a few vendors. Most of the equipments were "black-box" composed of dedicated hardware and software offered by the system vendors. These devices communicate with each other through standard communication protocols. In the evolvement of mobile communications network from 2G to 4G and of core network from softswitch to IMS, communication network elements (call control, media forwarding, services processing) have become more softwarized. However, most communications devices still adopt dedicated hardware for communications and the black box model. The mobile communications network still maintains the vertical architecture where all vendors "go in their own ways" and most devices are with dedicated usage without unified cloud infrastructure, thus failing to achieve resource-sharing.

Besides, operators have built dedicated core machine rooms, aggregating machine rooms and access machine rooms for corresponding systems of the mobile communications network. For a long time, there was no unified standard for mobile machine rooms and IDC machine rooms, which have been designed and built separately. There is much room for mobile machine rooms to transform and upgrade in terms of operating general purpose infrastructure. In the networking design oriented towards the future network, the unified planning of machine rooms also merits consideration at the same time.

The physical design of the future mobile communications network should have the following two basic features: one is "cloudization", that is, a mobile network with a focus on software will be carried at a unified cloud infrastructure; second, "more data center", that is, standardized data center will become the basic component of the future network.

The idea of NovoNet follows such a design principle: using TIC node as basic components, the network elements are deployed on-demand on the TIC nodes at the central, area and access layers, adopting fast configuration and adjustment through an unified management and orchestration system. All the above are connected to each other via layer-3 TIC nodes and high speed IP networks, forming the NovoNet in the future.

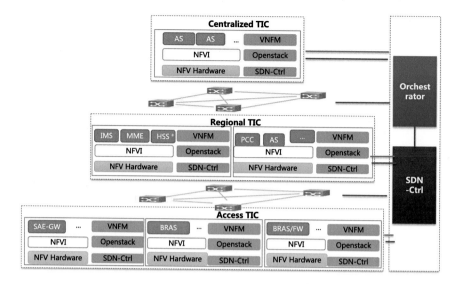

Fig. 1 NovoNet network architecture

The NovoNet network will be one featuring centralization, standardization and DC integration, as shown in Fig. 1. Centralization is reflected in network centralization and management centralization. Network element softwarization helps the centralized deployment and centralized network arrangement and orchestration promotes management centralization; standardization is mirrored in hardware with unified standards, virtual platforms with unified standards, and network element function with unified standards; DC integration is demonstrated in holistic planning of the layer-3 TIC architecture replacing traditional telecommunications data center and standardized TIC design. It can share hardware, transport and other resources with IDC.

1. Basic component of NovoNet: TIC

We defines TIC as the standardized infrastructure environment for the integrated deployment of telecommunications software, including the COTS general purpose server and hypervisor that meets the operator-grade quality, standardized machine room design and device networking requirements.

The high quality of TIC operating is mainly reflected as:

(1) High-reliability: high-reliability at the hardware layer, virtualization layer and virtual network function; system-level high-reliability between the layer-3 nodes.

(2) High-performance of the media forwarding element: accelerating technologies to support huge throughput, such as DPDK+OVS/ODP, SR-IOV; accelerating technologies for trans-coding and encryption, for instance, built-in chip accelerating module, board acceleration and accelerating pool.

(3) Highly efficient maintenance ability: acquisition of KPI (key performance indicators) data which is necessary for the operator's management at the infrastructure level; fast detection of the failures at hardware level, virtualization level, operation of virtual network function software, virtual network connection and cloud management platform (for example OpenStack); capability of fast restoration, rapid failure reporting and troubleshooting.

(4) Completed network management ability: virtual network function could be managed by OSS; basic resources that virtual network function relies on could be managed by OSS and NFVO in a coordinated way.

(5) Carrier-class physical isolation requirement in networking: physical isolation at the network management plane, infrastructure plane, service plane; all planes need to support the 1+1 redundancy of high speed network connection.

According to the operator's networking requirement, TIC will be categorized as central, area and access nodes to carry different types of network elements.

2. Management, arrangement and dispatching of NovoNet

A full set of management, arrangement and orchestration system is the core of NovoNet. In our view, NovoNet's management, arrangement and orchestration system is facilitated by comprehensively functional Orchestrator and SDN controller. Comprehensive Orchestrator boasts the ability of NFV's arrangement and SDN's uniform traffic dispatching. On one hand, Orchestrator, in partnership with NFV and VNFM, completes the network's management and arrangement, network service and life cycle management of the virtual network function and resources management; on the other hand, in combination with the SDN controller and OpenStack, it will achieve the interconnection of the virtual network function within TIC, involve the service chain on-demand, automatically deploy the interconnection between TIC nodes, and realize the mutual access of resources among TIC and resources integration.

Pioneer and Trailblazer

China Mobile has always been a fervent advocate and leader in promoting NFV and SDN. While deepening R&D, it spares no effort to mature the NFV and SDN industries with industrial partners.

China Mobile's technology staff have been enthusiastic in developing advanced technologies and are full of inspiration and innovation. As early as the beginning of 2010, China Mobile got down to research projects on DSN and C-RAN, introduced technologies like P2P, cloud computing, and started to research and test softwarization and visualization on general purpose hardware platforms.

With years of preparation, China Mobile proposed a core-network-cloudization featured system structure, which to a great extent resembled the later NFV concept/structure.

With NFV and SDN becoming the focus of the industry, China Mobile was enthusiastic in R&D, particularly with multiple SDOs and open source groups, aiming to mature NFV and SDN in joint force.

In 2013, China Mobile invited more than 20 vendors to China's first SDN testing and drove the establishment of the Carrier Grade SDN forum in the Open Networking Foundation.

In May 2014, China Mobile with its partners initiated the first NFV study item in 3GPP, i.e. virtualvirtualized network management project to research the key issues of NFV deployment.

China Mobile has been keen on industrial cooperation on a win-win basis. In 2012, it announced the first NFV white paper together with another 13 global operators, driving the establishment of ETSI NFV ISG, and kicked start the development of the NFV industry.

In September, 2014, as a major initiator, China Mobile founded an OPNFV project and served as director in the Linux Foundation, an influential open source group to study open source NFV platform at telecommunications level and to solve the problems in integrating open source platforms.

In 2015, China Mobile hosted the NFV Global Workshop twice to push the industry mature with industrial partners.

In early 2015, China Mobile announced the plan to build the first Open NFV lab in Asia, signed MOUs with 9 partners to set up the lab together, and performed testing and verification to provide the OPNFV system with an open international testing environment.

In March, 2015, China Mobile issued the *NovoNet 2020 Vision* on the Mobile World Congress, introducing the NovoNet brand to the industry, explaining its idea of development and the industrial demand of development, and integrating the industrial forces.

In April, 2015, China Mobile served as VP of the Open Networking Foundation and chair of the operator-class SDN working group.

In December, 2015, China Mobile proposed the idea of Open Orchestrator on the OPNFV Summit, demonstrated the prototype of Open-O, and invited players across the industry to join the open source R&D of Orchestrator.

While developing the industry, China Mobile constantly tries to implement and commercialize NFV/SDN as soon as possible.

In February, 2014, China Mobile showed vEPC- and vIMS-based end-to-end VoLTE virtualizedvirtualized system on the Mobile World Congress. In July, 2014, it showed the system prototype of NovoDC and NovoWAN on CommunicAsia. Afterwards, it demonstrated the NovoNet-based data center and WAN system prototype on the Mobile World Congress in March, 2015.

In September, 2014, China Mobile started to deploy NFV architecture of the New Call, New Message platform to integrate telecommunications commercial projects.

In October, 2014, China Mobile carried out the experimental work of core network cloudization (to apply NFV in vLMS and VoLTE) in four cities.

Industrial Development on Win-Win Basis

While developing the industry, China Mobile keenly felt that future communication networks must rely on the deep integration of CT and IT, which will motivate the industry to upgrade and transform to a higher level and propel network operators to undergo in-depth transformations.

From the perspective of network virtualization and NovoNet development, the existing telecommunication industrial chain will be broken with the entry of new IT players (such as open source vendors like Red Hat, VMware and hardware vendors like HP and Lenovo); the role of the existing system vendors may change as well, from equipment suppliers to software suppliers; new cloud management platforms and SDN controller suppliers are needed; competent integrators are also needed to provide packaged system products. This means a new chance for the CT and IT industries and operators to make choices and re-position. NovoNet is bound to rely on a brand-new industrial chain and at the same time give birth to a new industrial cooperation model.

First, industrial transformation takes shape in the form of the cooperation of CT and IT. Despite rivalry in certain areas, the two industries mutually reinforce each other for common development. The development of IT is based on the sound infrastructure and the capacity of openness provided by CT while CT needs to absorb new technologies and experience from IT for development. In this round of industrial upgrading, the CT industry needs more support from the IT industry in providing products demanded by communication businesses in terms of virtualization, cloudization and reliability. This is also a chance for the domestic IT industry to enhance strength and become an international player. It is likely that domestic IT businesses will gain strength, and a complete virtualized industrial chain with international competitiveness will take shape if we give full play to operators' guiding role through demand; if joint efforts are made by the entire industry to integrate technology solutions; if we make good use of the upgrading opportunities to develop software and hardware products with quality that satisfy the requirements of operators.

Second, open source cooperation has been a significant method of industrial chain development. Open source will effectively lower the threshold of the NovoNet industry and give birth to a lot of new players. Presently, open source groups are weak in supporting communication needs and domestic businesses are not influential enough; in the future, domestic businesses should join world-known open source groups (such as OpenStack, OPNFV, ODL, etc.) more keenly, strengthen international influence, improve availability of open source products, continue R&D work based on open source, and enable themselves to transform from open source to commercial products and services.

China Mobile will remain open, pro-cooperation and keen on reform. It will help guide industrial development, seek for a win-win scenario and together with industrial partners foster the bright future of Telecommunication 4.0.

Afterward

In June, 2015, I attended the seminar on IT application and new industrialization hosted by the Organization Department of the Central Committee of the CPC (ODCC) and co-hosted by the Ministry of Industry and Information Technology (MIIT). The seminar was held at Guanghua School of Management, Beijing University for three days and in Helsinki, the capital city of Finland for about three weeks. Those who attended the seminar included leaders of ODCC and MIIT, major leaders of certain provinces, cities, municipalities and local Commission of Economy and Informatization and Commission of Industry and Information Technology, and several leaders of central enterprises. This is a precious chance for discussion and study.

The seminar was held when Premier Li proposed the "Internet+" plan in his government work report during the Two Sessions, and MIIT initiated the "Made in China 2025" development strategy. As a result, "Internet+" and "Made in China 2025" became key topics of the seminar. In preparation for the seminar, I read through the book *Industry 4.0* on the flight from Beijing to Helsinki. The book was written by a team headed by a German, Ulrich Sendler and expounded on the upcoming fourth industrial revolution. Once published, it aroused a wave of heated discussion on the fourth industrial revolution across the world, making it a landmark work on Germany's industrial leadership.

Summer in Finland boasts clear skies and a warm breeze, particularly long days and short nights. The sun did not set until 10 pm and rose again at 3 am in the morning. Apart from intense study, exchanges, discussion and visiting people and places, I enjoyed a stroll by the lake at dusk and watching the sunset. The seminar lasted for a dozen days. Taking the seminar into consideration, particularly the thinking on informatization and new industrialization, I started to review the history and future of communication from a professional perspective. Probably inspired by the concept of Industry 4.0, I came to the idea of Telecommunication 4.0, which constitutes the topic of this book.

Back in China, I discussed the concept with colleagues at the China Mobile Communication Research Institute, and the idea was applauded by all. With in-depth discussion, we came to the conclusion that Telecommunication 4.0 was rich in connotation and forming an ever clearer theoretical system. After exchanges

© Springer Nature Singapore Pte Ltd. 2018

Z. Li, *Telecommunication 4.0*, DOI 10.1007/978-981-10-6301-5 with industrial leaders and experts, we agreed that it was a good notion and that a book should be written in order to spread and share it. Thus my colleagues and I devoted plenty of spare time to writing the book *Telecommunication 4.0: Reinvention of communication network.*

My acknowledgment first goes to Wang Xiaoyun, president and Yang Zhiqiang, vice president of the China Mobile Communication Research Institute, and a team of young specialists, including: Duan Xiaodong, Wang Shuaiyu, Zhang Hao, Yang Jing, Fu Qiao, Ling Jinglei, Lu Lu, Li Chen, Deng Hui, Xue Haiqiang and Wang Lu. They have contributed prominent research achievements and are the co-authors of this book. Thus this book has been accomplished with our joint efforts.

I am grateful to my three respectable former superiors and experts who prefaced the book. The first one is Mr. Wu Jichuan, former minister of Ministry of Posts and Telecommunications and Ministry of Information Industry, who has been devoted to posts and telecommunications industry. He has led China's communication industry to stride into Communication 3.0 from Communication 1.0, a great leap in China's communication industry. The second one is Mr. Wu Hequan, fellow of the Chinese Academy of Engineering and president of the Internet Society of China. He has been an authoritative expert active in the information and communication industry home and abroad. The third one is Mr. Zhao Houlin, the present secretary-general of the International Telecommunication Union. He is the first Chinese person to hold this position and one of the Chinese citizens in senior posts at UN agencies. They emphasized the significance of Telecommunication 4.0 and its influence on the whole information and communication industry in their prefaces.

The sixth chapter is made up of essays contributed by relevant institutions and experts. Authors include Zou Zhilei, president of the carrier BG of Huawei Technologies Co Ltd, Werner Schaefer, HP global vice president, Ding Lei, founder, chairman of the board and CEO of NetEase, Tian Shuoning, president of China Broadband Capital Partners and president of AsiaInfo and Huang Xiaoqing, founder of Cloudminds Inc. They reviewed Telecommunication 4.0 from various perspectives and shared their understanding of opportunities and challenges brought about by Telecommunication 4.0. Herein I'd like to extend my sincere gratitude to them.

My gratitude goes to Mr. Wang Jianzhou and Xi Guohua, former presidents of the China Mobile Communications Corporation, Mr. Shang Bing, the present president and Mr. Li Yue, chief executive. I could not have decided to write this book without their understanding and support.

Finally, my thanks go to my beloved wife Xiangjun and daughter Qinglan for their understanding and care when I was conceiving and writing this book.

Closing the book and pondering, I recall the time 25 years ago. Then I was a young associate professor at UESTC, a "Chinese Ph.D. of prominent contribution" accredited by the former State Education Commission and Academic Degree Commission of the State Council and a winner of the Science and Technology Award for Chinese Youth awarded by the China Association for Science and Technology. At that time, in my prime and full of ambition, I left for the US on an academic visit. I attended the academic conference held by the Institute of Electrical and Electronic Engineers (IEEE), and visited the University of California in San Diego, Los Angeles, and Berkeley as well as Stanford University. I was invited to give lectures and undertake academic exchanges. The most unforgettable was a visit to the Jet Propulsion Laboratory at the California Institute of Technology, where Qian Xuesen worked. I was overwhelmed with excitement. I made up my mind to contribute to China's science and technology endeavors just like him.

Today, looking at this book on Telecommunication 4.0, I cannot help but wonder if it is a historic chance for a Chinese person to expound on the idea of future communication, which predicts that China is moving from a major communication country to a communication power, that China is on the way to realize the grand goal of becoming powerful through networks—a notion proposed by President Xi.

If I had the chance to help it, I would never regret dying the moment this dream came true.

Li Zhengmao
December, 22, 2015